Model Railroad Elec

Rutger Friberg

Allt om Hobby

Model Railroad Electronics 3

Author: Rutger Friberg

Original title: Mer Elektronik för modelljärnvägen, 1989 ISBN 91-85496-32-4

By the same author in the English language:
Model Railroad Electronics, 1993. ISBN 91-85496-73-1
Model Railroad Electronics, 1994, ISBN 91-85496-82-0

By the same author in the Swedish language:
Elektronik för modelljärnvägen, 1st edition, 1986. ISBN 91-85496-24-3
Elektronik för modelljärnvägen, 3rd edition, 1989. ISBN 91-85496-35-9
Digitaltåg, 1988. ISBN 91-85496-29-4
Mer elektronik på modelljärnvägen, 1989. ISBN 91-85496-32-4
Elektronik för modelljärnvägen 4, 1994, ISBN 91-85496-78-2
Elektronik för modelljärnvägen 5, 1995, ISBN 91-85496-80-4

Cover photo: Jens Dahlström
Layout and production: Freddy Stenbom

First published in the English language in 1996.

ISBN 91-85496-91-X

Publisher: Allt om Hobby AB, Box 90 133, S-120 21 Stockholm, Sweden.
Phone +46 8999 333 Fax +46 899 8866
e-mail: freddy.stenbom@hobby.se Information on Internet: http://www.hobby.se

Printed in Sweden by Gummessons Tryckeri AB, Falköping

Introduction

The first book in this series was never expected to spawn sequels and labeling it with leading phrases such as "volume one" was carefully avoided. The odds for success were stacked against it from the beginning - demand for a book on model railroad electronics, written in Swedish couldn't be great. I would like to take this opportunity to thank the publisher for taking the gamble. Happily, many were obviously interested in this sector of the hobby.

I've assembled a new collection of construction projects and assorted tips. Most are model railroad related and independent of modelling scale, although size restrictions in Z scale may limit the usefulness of some. Included are several alternating current, Märklin-specific projects.

In the first book, I tried to be thorough concerning soldering methods, selection of tools and theoretical background. The construction projects were simple in the beginning, and have increased thereafter in both functionality and degree of difficulty.

In this book, you'll find the same general arrangement but this time I've raised my sights a bit. We'll be looking at an expanded array of tools and instruments. The accent will shift to function and electrical schematics and away from construction methods. The latter is well covered in the first book.

Several projects are collected under the common theme of "program-control". You don't need to know a thing about computers or own one or even have the slightest interest in them; just the same, we'll be developing advanced sequences that when presented on the layout, are guaranteed to impress the crowds. A limited amount of basic theory has also been included in an attempt to answer the common question, "How does it work?".

Some readers may be interested in electronic projects unrelated to model railroading. I hope that they'll also be satisfied with the book's contents.

Finally, I'd like to thank my family for putting up with a home filled with experimental projects, prototypes and manuscript sheets, and for also contributing many valuable project ideas.

Rutger Friberg

P.S. Special thanks to Professor Thomas Catherall, Stanley R Ames and Richard H Lord who added value and comments to the book. D.S.

TOOLS AND EQUIPMENT

In the first book I tried to be thorough in descriping tool requirements for the beginner. With this fundamental knowledge as background, I'll now make a more generalized summary of the basic requirements as well as looking forward to the expansion of the hobby workshop with a practical economic view.

From a hobby standpoint, if you are beginning with just "two empty hands", don't rush out and buy up the inventory of your local hardware store. Your investment will be too large and misplaced - what you'll need to get started is a rather small but well-chosen basic kit. Limitations that arise usually aren't due to lack of tools but rather to a lack of knowledge, experience and the patience to puzzle together small objects.

When you begin collecting tools for your first electronics projects the first thing you'll need is a screwdriver set of small screwdrivers - some should fit cross-slotted (Philips) screws. Furthermore, you'll need pliers and wire cutters suitable for bending and clipping component leads and stripping fine gauge wire. We'll be stripping wire with

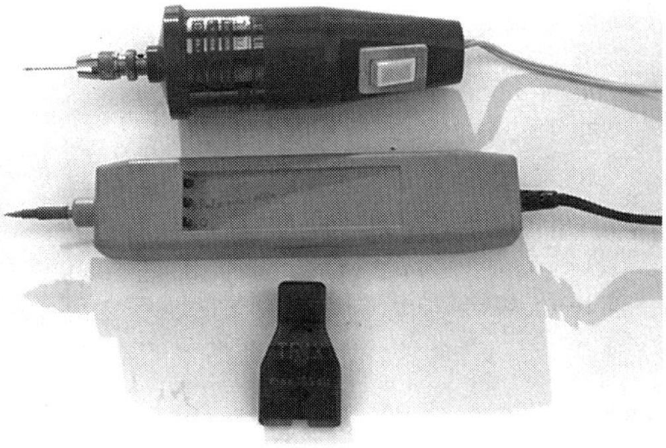

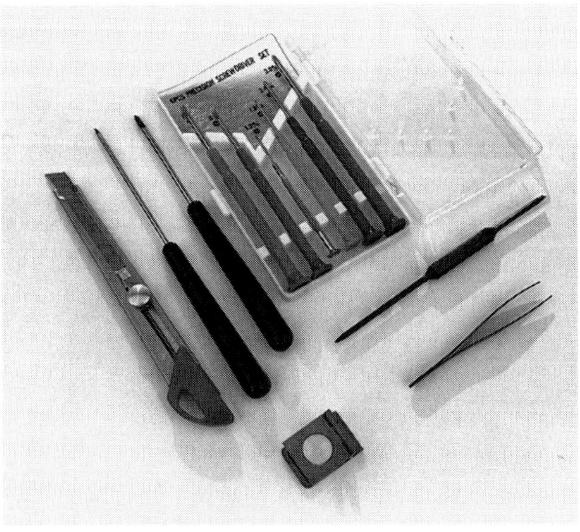

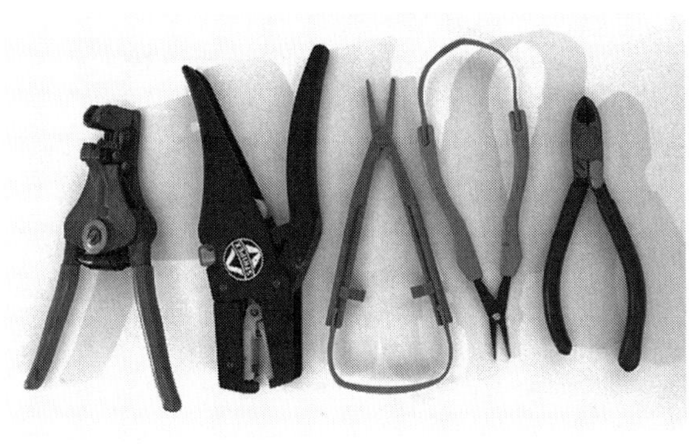

diameters of 0.25 mm to 1.0 mm. It is unnecessary as well as expensive to purchase heavy duty wire strippers. It's important that the clippers are small and easy to use. Larger sizes should be used in the garage and smaller ones in the hobby room. Both small screwdrivers and wire clippers can be purchased together in beginners' kits. Basic soldering pencil kits are also available. Earlier generation soldering irons can be hung on the wall as a conversation piece.

You'll need a soldering pencil with moderate wattage, yet small enough for close quarters and printed circuit card work. Solder should be as thin as 1 mm wire and with a resin core. At the beginners' phase, you can top off your toolbox

assortment with a battery powered, miniature drill. These are available from hobby supply firms and suitable drill bit sizes are in the 0.8 to 2 mm range. As long as you drill without a stand and drill infrequently, common carbon steel bits will suffice. Later, you can switch to higher quality bits. When you're beginning to feel really professional, make the change to hard steel bits. These are almost impossible to wear out but extremely sensitive for angled and lateral forces - they'll snap immediately with the least amount of object movement. It is unnecessary to pay several dollars for a single bit to discover this phenomenon.

MEASURING DEVICES

Many hobbyists are tempted to use test pens with light emitting diodes (LEDs). These instruments often resemble two nails, linked with a cable. I do not recommend these for beginners except as a complement to a multimeter - which should be included in the basic investment. Beginners' multimeters are priced from $10 to $20. A suitable multimeter should be capable of measuring both alternating and direct current voltages of at least 500 volts. Several scale selections are valuable when you need to differentiate between 4.5 volts and 5 volts, for example. Furthermore, the instrument should be capable of making direct current measurements within the milliampere range as well as maximum currents in the 5-10 ampere range. The multimeter should feature an Ohm scale for testing various resistances (resistors). These are not luxury options and should be featured on instruments in the stated price class. At the beginning, avoid digital instruments with numeric displays. For both price and readability reasons, the beginner's multimeter should be analog - with a needle pointer!

Compared to this investment, all other equipment and materials are of the cut-rate type. As far as project materials are concerned, you'll be buying them as needs arise. Of course, it's understood that your basic supplies include thin copper wire, electrical tape, thin brass wire, glue and the like.

MORE TOOLS

As hobby finances justify expanding to your tool kit, you can begin by raising the quality of your hand tools, if these aren't already of sufficient quality.

It's easiest to test this on a pair of wire clippers. Close them and hold them up to the light. The more "peep holes" there are, the less suitable they are for their tasks. Here is an area where quality pays.

Complement your kit with high quality tweezers. Next, buy a small vise. There are vises available with suction cups and a wide array of positioning variations. You can forget them at this stage. Prices can exceed $50. Your needs can be covered for around $10. One item that is quite practical is a so called "third hand". Several alligator clips and a magnifying glass are mounted on a stand. These sell for approximately $10.

POWER TOOLS

We'll ignore all of those nice lathes and machines that the mechanic requires and concentrate instead on the needs of the electronics hobbiest. I'm going to stick out my neck and make the following suggestion. If you feel that electronics is enjoyable, instructive and "your song", an investment in better soldering equipment can be in order. If that's the case, you'll need a soldering station. With one of these, you can switch tips as required and regulate heating wattage.

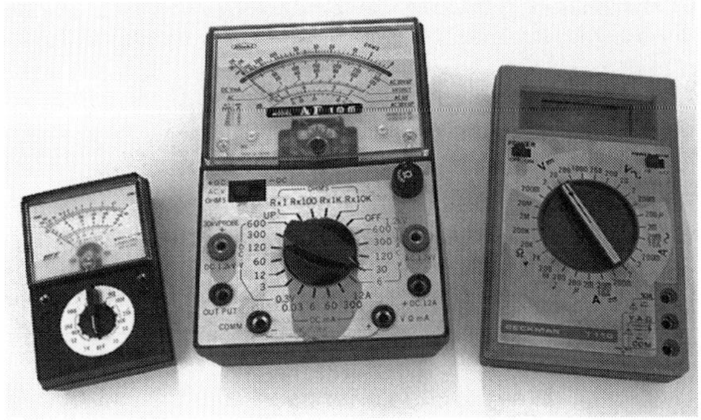

MINI-TIP

Voltage tester

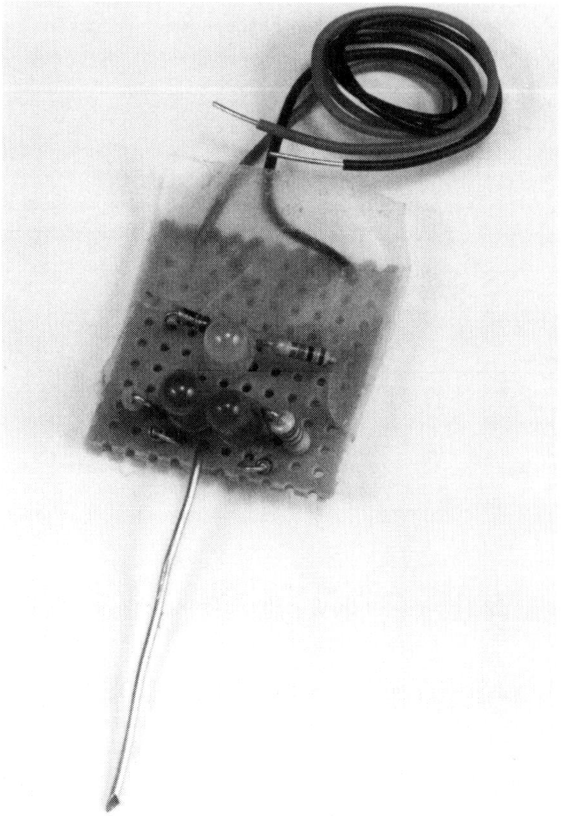

This is a simple voltage tester which is handy in case you don't have a more professional multimeter. Most simple testers of this kind will only via a lit LED show that a plus voltage is present. If you test a ground connection you can't be sure if it is connected to the central ground or if the point is floating.

With our tester the green light indicates that ground is present (common ground is necessary). The yellow LED is lit when the tester is ready to go. If this LED is the only one lit that is the sign of floating test point or a voltage in the intermediate level 3-9 volts.

Depending on scale you might want to test 8-12 volts instead of the 12-16 volts. Then change the Voltage converter to 7808 and the Zener diodes to value 4.7 volts. And the current limiting resistor of 1 kilo ohm should be lowered to 680 ohms.

For this latter design variant you may also use a 9 volt battery instead, in order to make it easier to build and quicker to use. Then you can omit the four supply diodes 1N4148, the capacitors, and the voltage converter. Battery plus should point towards the resistor with the highest value (680Ω). Don't forget to include the circuit breaker since you otherwise will soon have an empty battery.

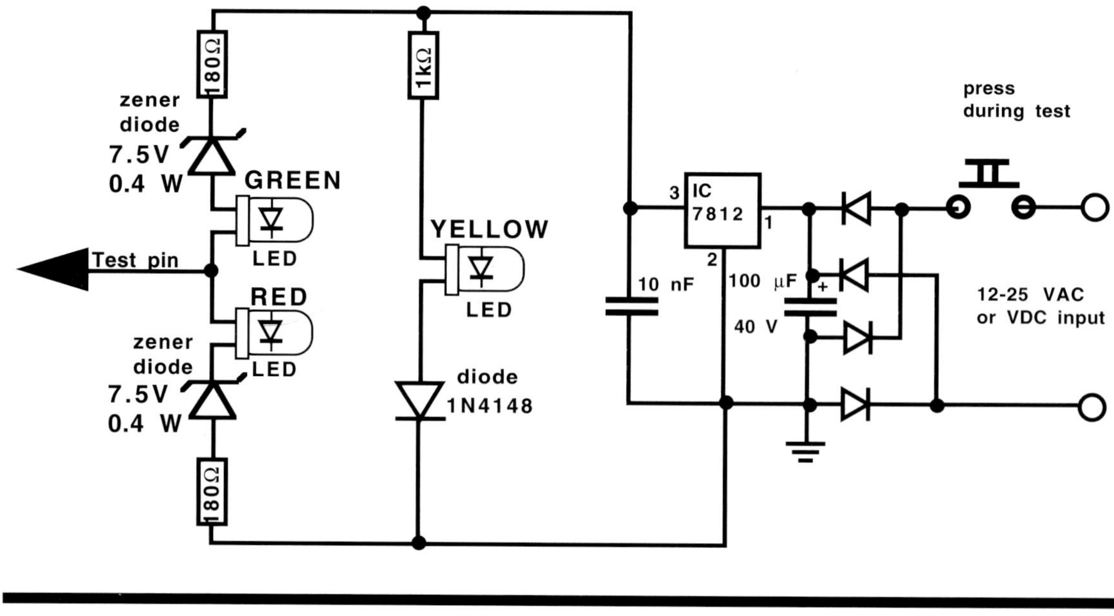

Soldering jobs will be more fun and the results will be better. A suitable soldering station costs around $100.

The next addition can be a bench drill for circuit boards. They feature stands and work with the previously mentioned hard metal drill bits. My own drill is an Amini and manufactured by Enomoto Kogyo in Hammamatsu, Japan. I purchased it in Sweden a number of years ago for about $180 and have since put it through the paces without a single problem. If you manage to locate one of these, I recommend them highly. They resemble drill presses in miniature.

A final suggestion regarding saws. After sawing several circuit boards, hand sawing becomes torturous. I recommend purchasing a small band saw. My economy model, Black & Decker has served me without fail for several years and has sawed a good 2,000 circuit boards. Talk about value...

MORE INSTRUMENTS

This is a section without end and potential investment size is unlimited. For most hobbyists' financial resources, we are now passing the boundary separating the hobby area from the professional, where only the best is good enough. Because of this, I'll limit my suggestions to a single item. My second instrument investment was in a simple and robust oscilloscope. I have an American made Gould Advance with rather modest performance capabilities and

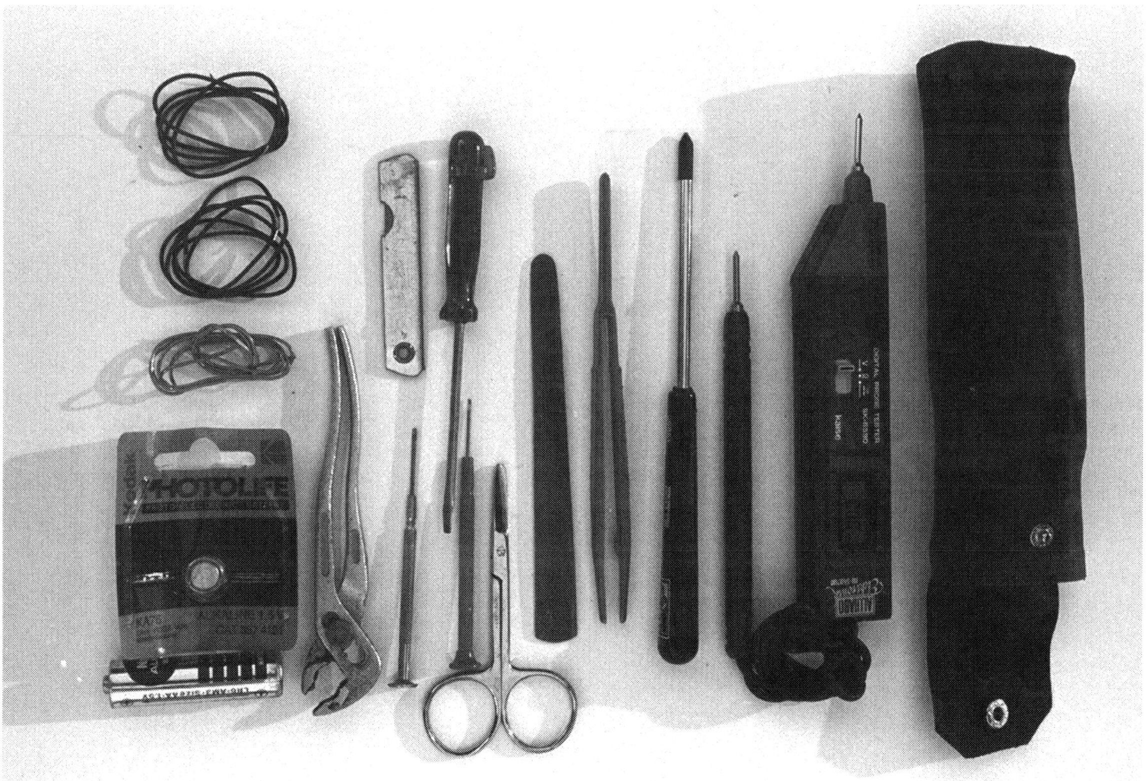

POCKET TOOLBOX

An emergency kit is invaluable at many instances. You never bring the big toolbox and therefor it is lightly not to have any tool at hand outside the hobby corner. Over the years I have gathered a separate small kit with tools for events like a train demo at a friends place, or at the local MR-club.

A small and handy testing tool for current, resistance and voltage (AC and DC) dominates the space. Screwdrivers,scissors, a small knife and a poly-grip are also included. A penlight cell, a piece of wire and some thin solder tin add up to this pocket toolbox.

at times, I'm forced to borrow a more advanced model (faster with more memory) for certain, "difficult cases". At present, there are several good brands on the market and several of them are bargain priced models from Japan. You can easily spend several hundred dollars for a new oscilloscope. It can be worthwhile to check the classified ads for a used instrument. If you have a friend well-versed in the mysteries of electronics, take him with you when you

This is a typical first prototype of a project in this book. Behind in the process is the wish list of needed functions. And after that some sketching on possible solutions were done, often including rough value calculating. The most often used calculations are those involving Ohm's and Watt's laws.

At this stage there are two main ways of easy making prototypes. One route is called wiring and gets use of a smart rotating wiring tool and a board just with high density of holes. The other one is used here. On this Protoboard you can mount all thru-hole components and then just interconnect them via short single wires with de-isolated ends. Underneath the holes there are electrical connectors (main area have North to South connectors and the more narrow top area has East to West connectors) which operate by friction and no soldering at this stage is needed.

Here is a close-up of the above mentioned Protoboard. As shown most IC-pins will fit directly and you can spread out components enough to make the module easy to test and review. It is also important to try to assign different colors to different tasks, although it doesn't show here of course. I use black wires to ground, red to plus feed and orange or yellow to AC. That certainly simplifies eventual fault finding.

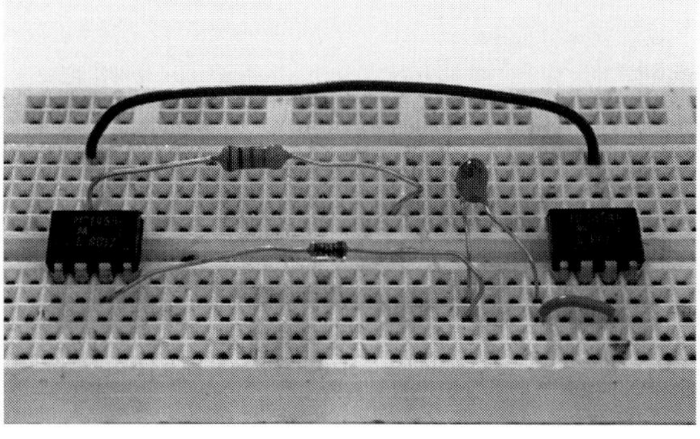

shop. Remember, the only tube in the casing should be the picture tube (cathode-ray tube). If it's not, the oscilloscope is too old. Beware of "burned-in" curves on the screen. This is an indication that the oscilloscope has been on for a long period of time, with the same strong light ray and the same signal image. Try to locate a manual. They are good introductions to oscilloscope use. I may return in a later context with more tips in this area.

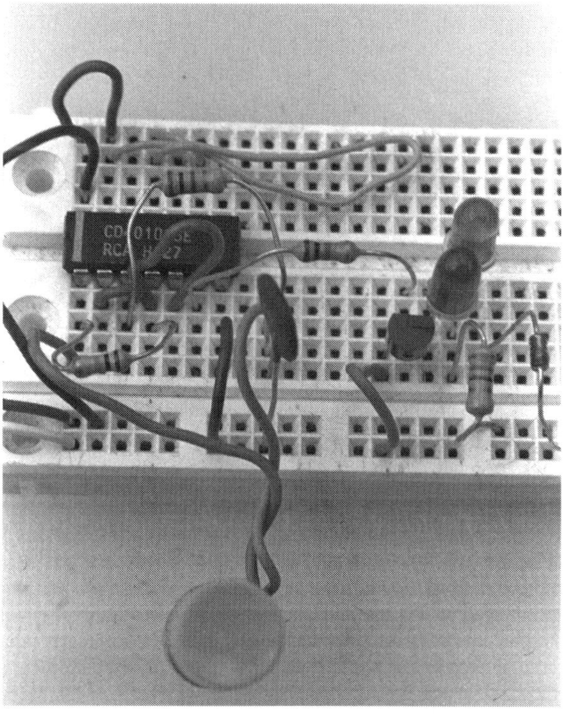

If you want to adjust the flashing home signal for fit the standards of your own layout — the flashing frequency or logic input level — then a temporary module on a Protoboard is a perfect reference. This allows you to change components values or add an extra gate "in a second" and no soldering is needed. On this prototype the input from the track is simulated through a button switch and a resistor, generating either a logic "low" or a "high".

CONSTRUCTION PROJECTS SECTION

Let the signals flash

The signals on our layouts are usually either static red or green. Just as on the prototype, they're easier to detect if they flash. In that it features diodes, the signal in this project can be powered with either AC or DC.

The signal is controlled without relays and flashes green as long as the A-input is grounded. This can be done with either a switch or a vacant relay contact pair. Connect Point A directly to its output on the control port, grounding it to the layout's common ground.

To use the unit without the flash feature, ground Point B. Non-flashing signals do not require the four NOR gates (an IC4001) that make up the flash oscillator.

This way of making signal control – low logic input means "stop" and high input means "go" – is fairly common. Most track blocking system use this logic for interfacing between blocks. Therefor you may well use it as an add-on. Don´t forget in such a case to secure common signal ground. If by any reason you get the wrong signal lighting, then simply switch wires between the red and green lights.

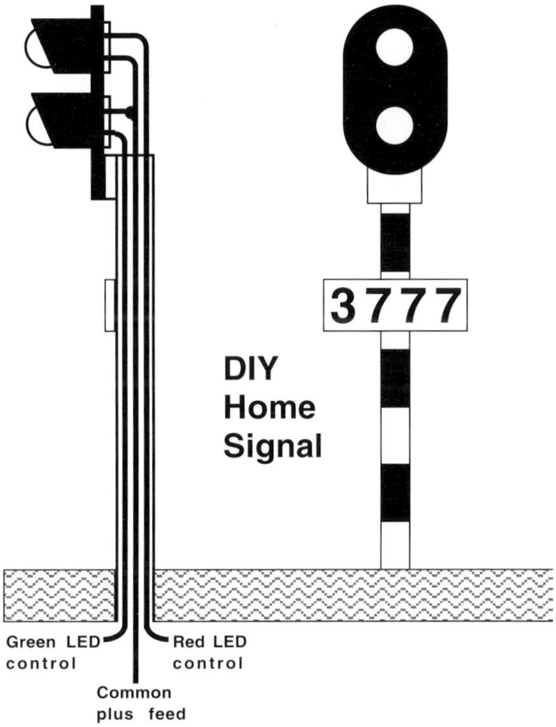

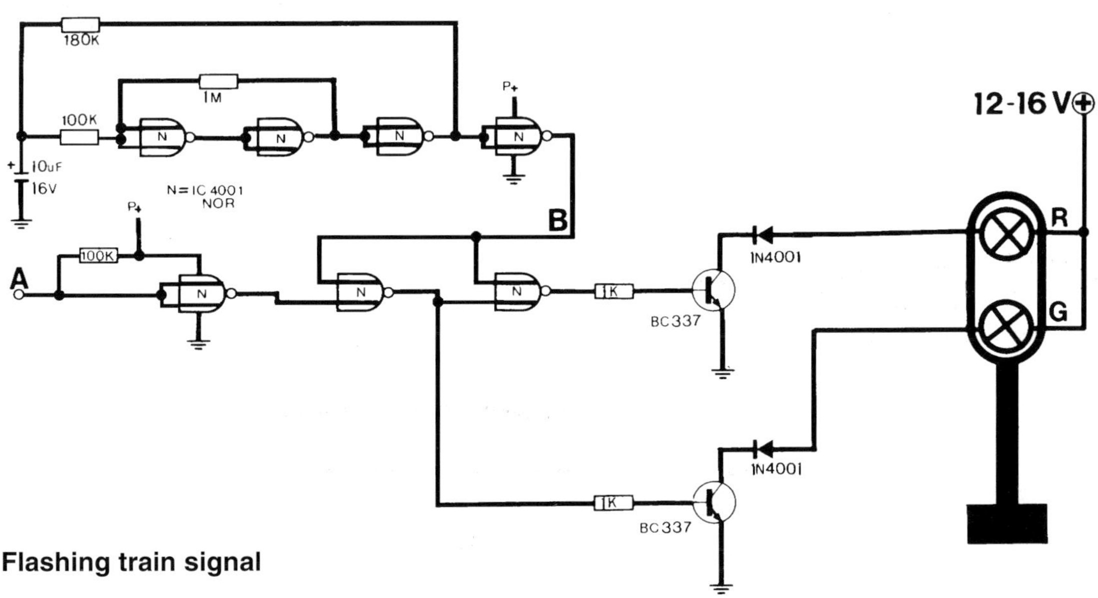

Flashing train signal

The light emitting diode - a lamp without limits

The light emitting diode, or LED, is on its way to becoming the most common component on the model railroad layout. They have many advantages over light bulbs. One of these is that they almost never wear out. Common miniature bulbs will probably need replacement after a few hundred hours. This can be a complicated operation if the lamp is soldered in a difficult place, such as in a signal lamp housing.

There are LEDs that cover almost all of our application and size requirements. Some have built-in flashing capabilities and others can vary color depending on current flow direction. There are LEDs with extra brilliance, almost like miniature headlights. The 1 mm diodes are even suitable for N and Z scale.

A few things are important to remember when connecting LEDs. A standard LED only lights in a single connection direction (see the figure) and because of this, there are visible differences between the anode (a) and the cathode (c). Furthermore, they can handle only a limited amount of current and must therefore be complemented with series resistors with values varying with current input (a chart with a selection of values is provided). When higher AC voltages are used, such as line voltages, common resistors are insufficient. In these cases, capacitors are used. Capacitor circuits are quite practical at times but should only be used by those having the necessary electrical skills.

LEDs are available in red, green, yellow and orange. Blue LEDs should be on the market within the foreseeable future.

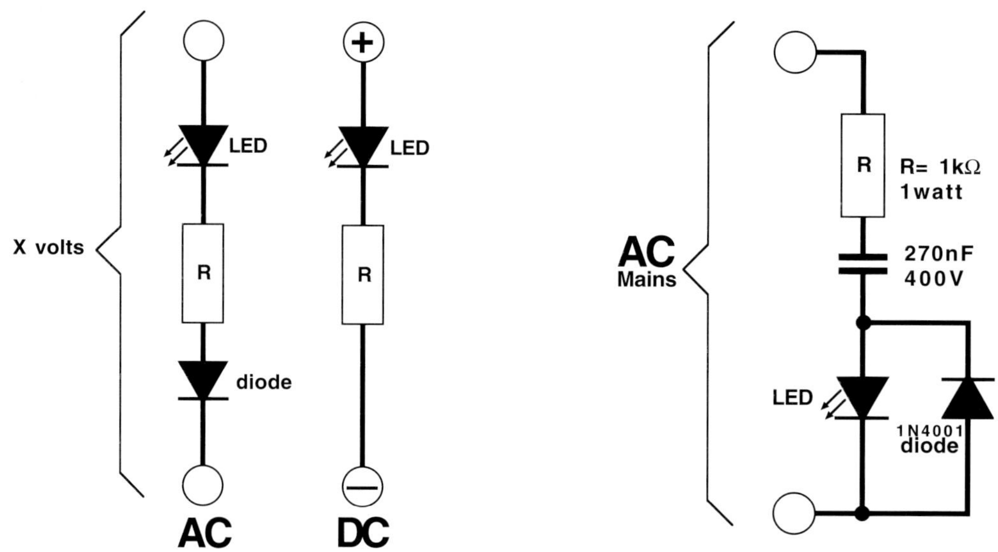

The Current Limiting Resistor (R):
How to select correct value

Power volts	for DC operation	for AC operation
6	680Ω	330Ω
12	1.0kΩ	470Ω
18	1.8kΩ	820Ω
24	2.2kΩ	1.0kΩ
30	2.7kΩ	1.5kΩ
40	3.9kΩ	2.0kΩ

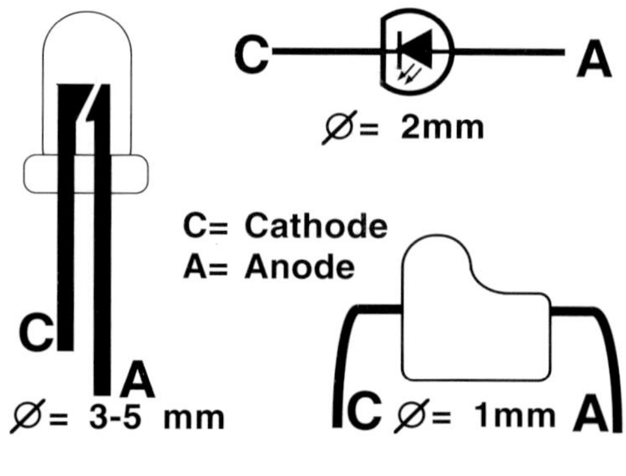

Railroad signals with one-button control

This circuit has several attractive characteristics. It is simple and inexpensive to build, and is a suitable project for those with little construction experience. Furthermore, it incorporates so few components that a circuit board is not necessary. This control unit can be built around IC 1 (40106) or even better, an IC socket. The IC can be inserted after the soldering work is finished and inspected.

ONE BUTTON FOR TWO FUNCTIONS

Two buttons are normally used for controlling signals' red and green lights. In this circuit, we'll save one of these for another project and let the remaining button switch alternate between red and green. Each time the button is released, the last selected color is retained. The control unit is actually a small bistable circuit, built with two gates - Schmitt trigger inverters in the six-gate IC 40106 (see the figure).

If you build this control unit, the gates shown in the figure are suitable. Pin connections are provided to erase any doubt as to how to assemble the project. Using the figure as a connection guide, you will be left with four unconnected IC pins. These inputs must be connected to a logic level (read more on this in the tips section). I would suggest soldering pins 5, 9, 11 and 13 to ground.
By now you have probably discovered that by using a single IC, three complete control units can be built and control just as many signals. Furthermore, it isn't necessary to use a push-button switch for control. We can let trains regulate the signals with a reed switch between the rails and magnets under the locomotives. Relay contacts can also trigger signal switching.

THE MULTIPLE VOLTAGE ALTERNATIVE

According to the schematic, input voltage should be 12-16 VDC. This is appropriate for the application for which this circuit was designed. In the figure for IC 40106, we can see that the IC can be powered with as little as three volts. In Z scale, for example, we may want to use 7 VDC. This isn't a problem. Just substitute the 1 kilo Ohm (kΩ) LED series resistor with one with a value of 470 Ω. With an input voltage of 10 volts, 680 Ω is a suitable value.

The control unit is powered with direct current and can be connected to a battery. You can, of course, connect the signal system to the accessory outlet on the same transformer as used for train control. This is an alternating current (AC) output meant for such accessories as lighting, switches, etc. The power unit (see the figure) was designed for just this type of connection. Connect it between the transformer and the signal control unit. It is capable of powering six signals without a heat sink and fifteen signals when IC 2 is mounted on a heat sink. Note that the IC's built-in heat sink is connected to the middle lead (and ground). This is only an important issue if the IC shares a heat sink with another IC - it could lead to short circuits. IC

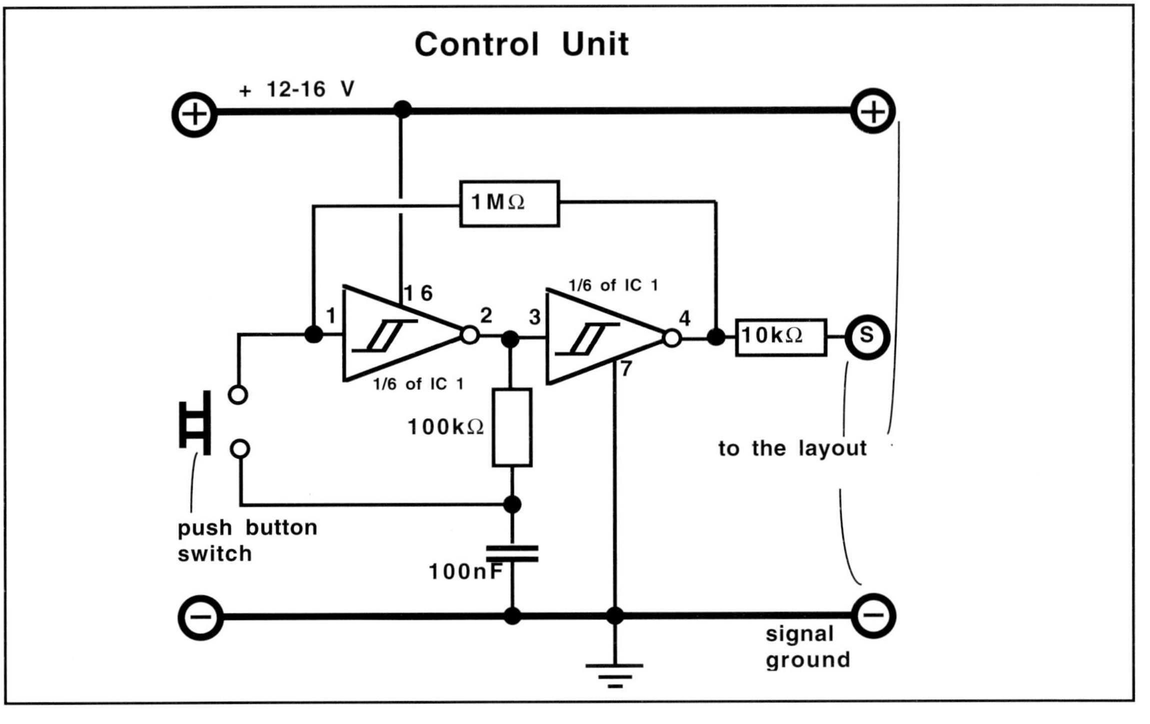

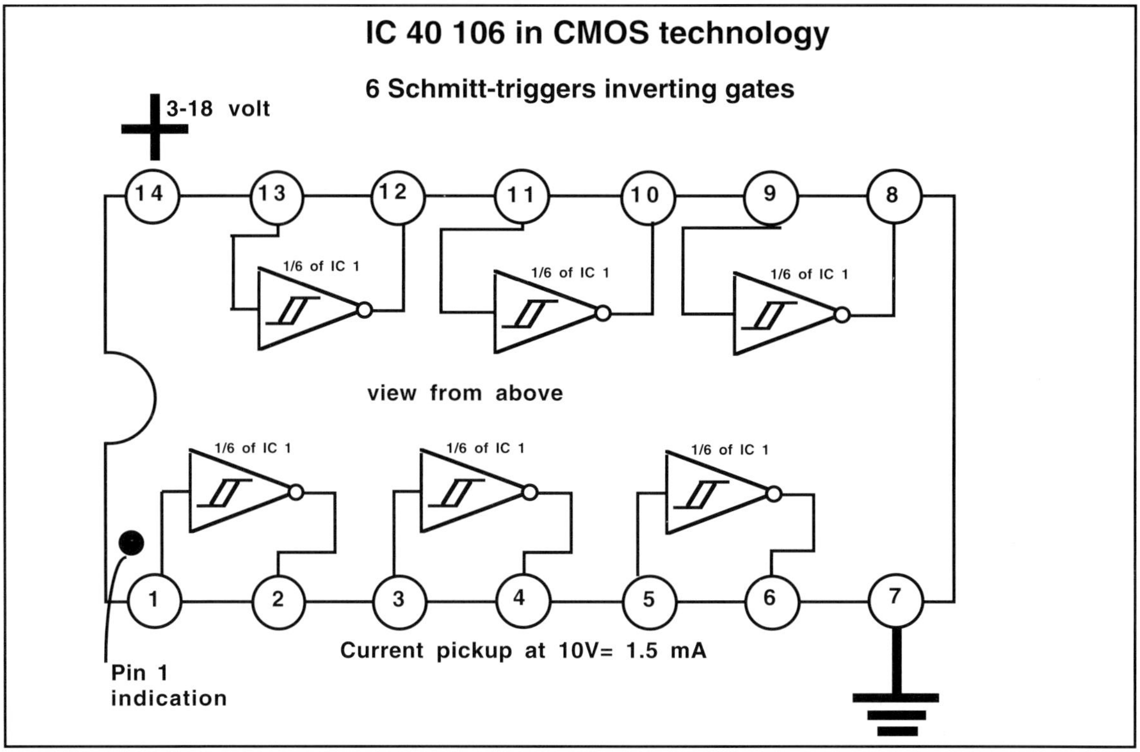

2 has a TO220 case that can handle up to one ampere. The same is true for the rectifier bridge consisting of four 1N4001 diodes. A ready-made bridge could be used if you have one in stock, but one of the basic goals of this book is to use as many basic components as possible.

A SIGNAL MAST WITH FIVE COMPONENTS

Four semi-conductors and a resistor is all that is needed for this signal. Of course you'll also have to construct the mast and lamp housings. A simple way to build them, besides using brass tubing, is to use circuit board laminate. You may find these circuit board signals in hobby shops as inexpensive kits if you don't have the resources to make them yourself.

The accompanying figure illustrates how to connect them. It's almost a luxury to use circuit board under the layout for just three components. I use small brass nails attached directly to the benchwork as connection points for wires and components. The resistors can be connected in any direction but the semi-conducters must be connected as shown. The transistors' base lead is connected to the control unit's S-input. All plus and ground leads can be connected together for common feed from the power unit. Finally, two little tips... The first is that the LEDs anodes are connected together, that's to say, turned towards each other. The other is that the red and green must not change positions with each other. They are connected in different ways and the two types differ a bit in regard to voltages - a characteristic that we will take advantage of.

Perhaps it is worth mentioning that this circuit was not

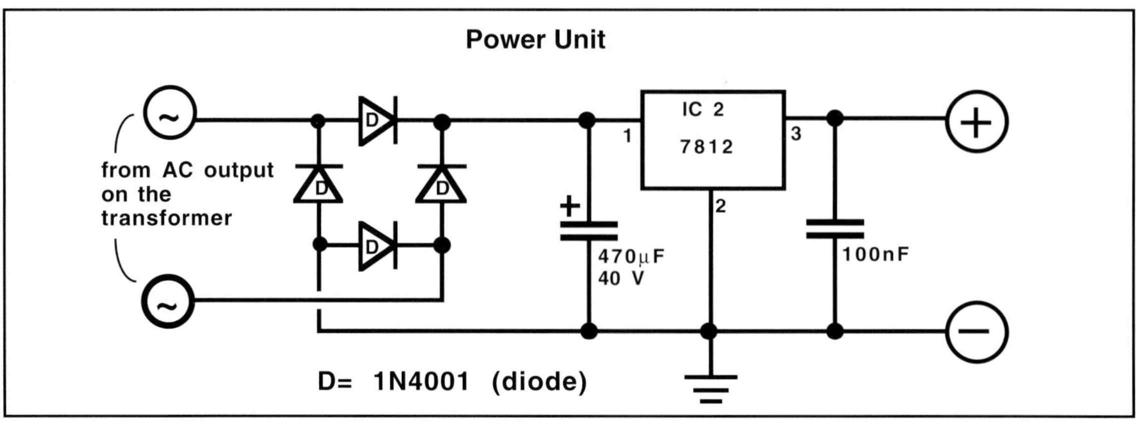

designed for train control. To the contrary, as noted earlier, we can let the trains control the signals. Those of you more skilled in electronics can easily add train control by complementing the control unit with an additional inverter at lead 4 that controls an power transistor and a relay. There are several examples of this “servo-connection”. See the Mini-Tip featuring a time relay.

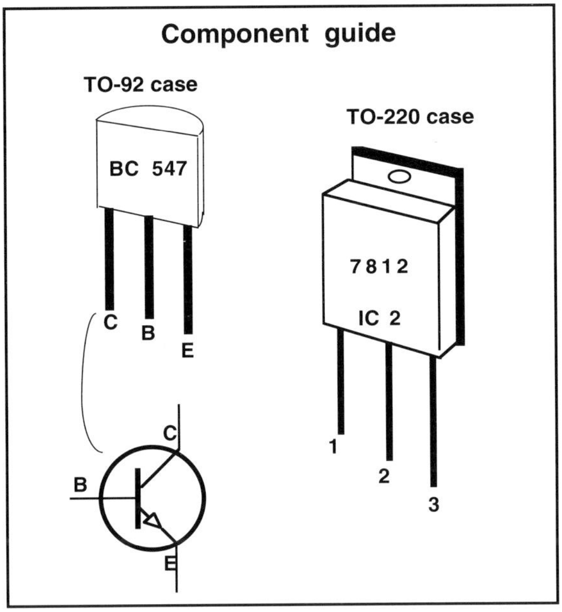

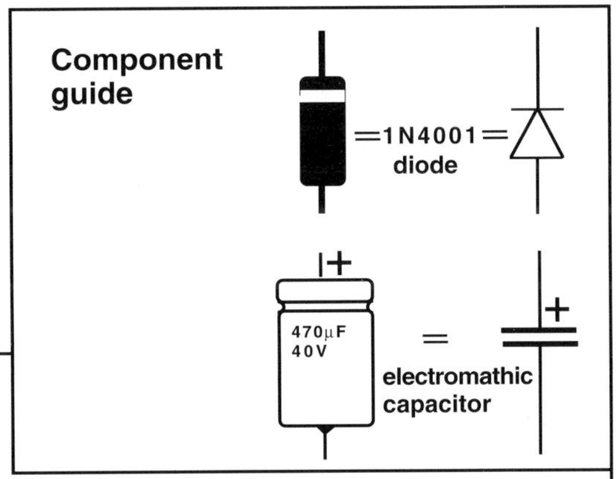

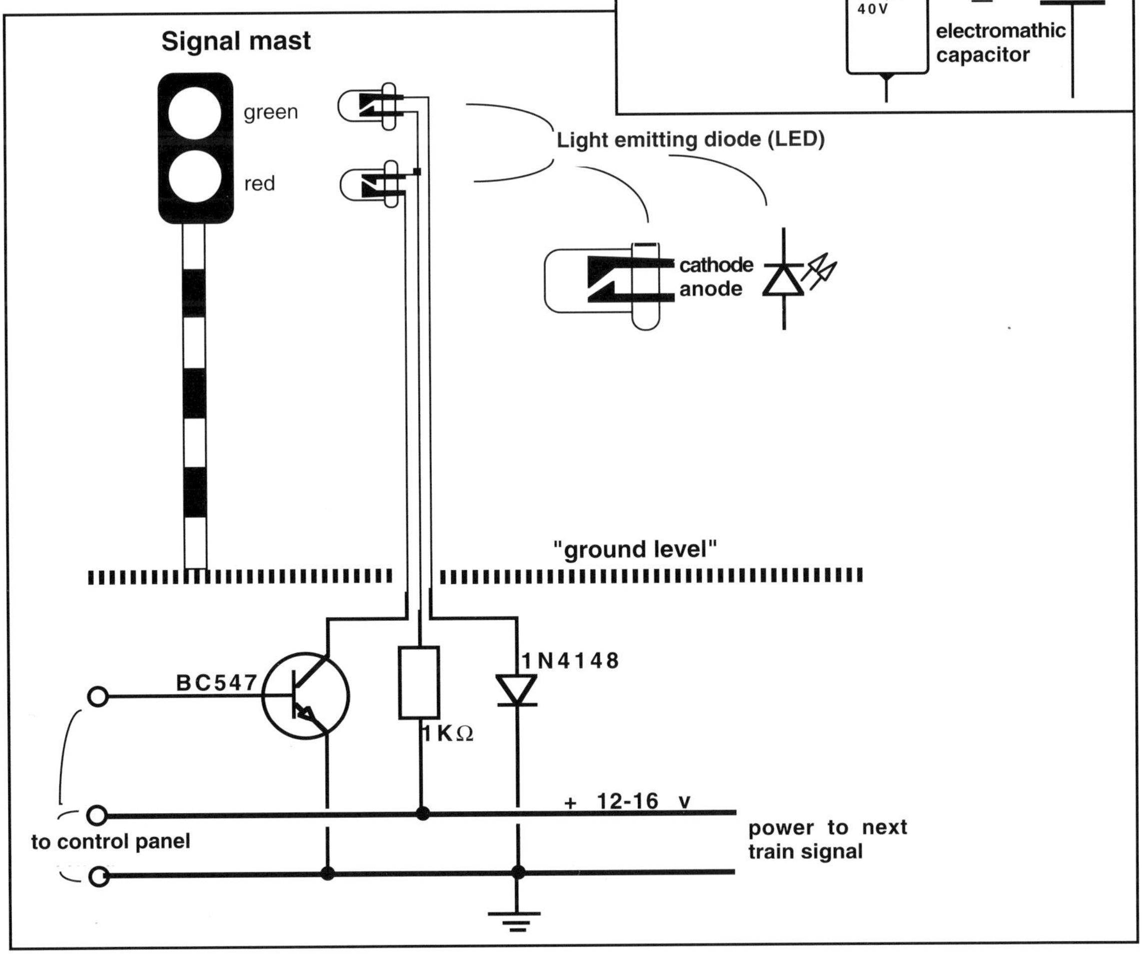

A controller for slow starts and stops

The common controller lacks features for really soft locomotive acceleration and braking. This circuit is a variation of a pulsed unit that was first presented by Peter Thorne in "Model Railroaders Projects" in the beginning of the '70s. Refinements have also appeared in the German model railroad press.

We'll be building a unit with present-day, common components readily available in electronics stores. It consists of modules that can be inter-connected and used as required. The controller is connected between the transformer and the rails. No alterations to the locomotive are needed. Dependent upon your power source (DC or AC), either module F1 or F2 is used. To change direction of travel on a Märklin locomotive, a 24 volt supply is required. This is available at lead W.

All speed regulation takes place in module B where acceleration and braking are adjusted with a trimming potentiometer for each function. A switch starts and stops the locomotive and the desired running speed is regulated with the master controller (a trimming potentiometer in module B).

Module D's job is to create a short delay between powering-up to avoid jackrabbit starts.

Running current is regulated in module E by a heat sink equipped, power transistor. For larger HO locomotives, exchange the 0.5 Ω resistor for a 0.3 Ω resistor.

An overload indicator is also included in module E. This can be replaced by a relay (12 V) and a protector diode to cut power. The transformer section should also feature a safety function. If you use a transformer meant for model railroad use, this is a standard feature.

The controller unit provides acceleration and braking times of up to one minute and engine speed is barely detectable at start up.

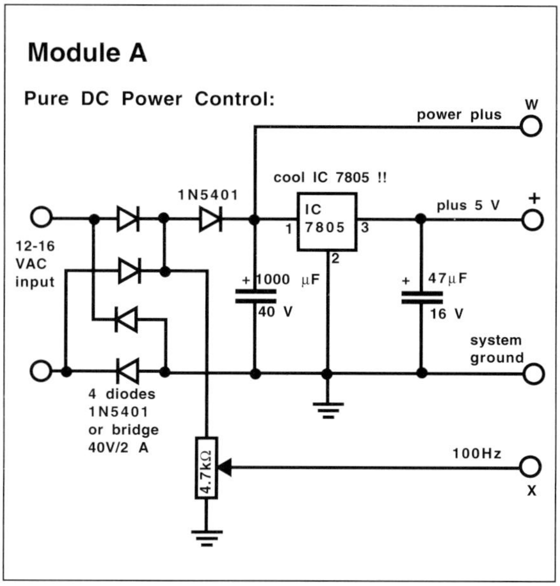

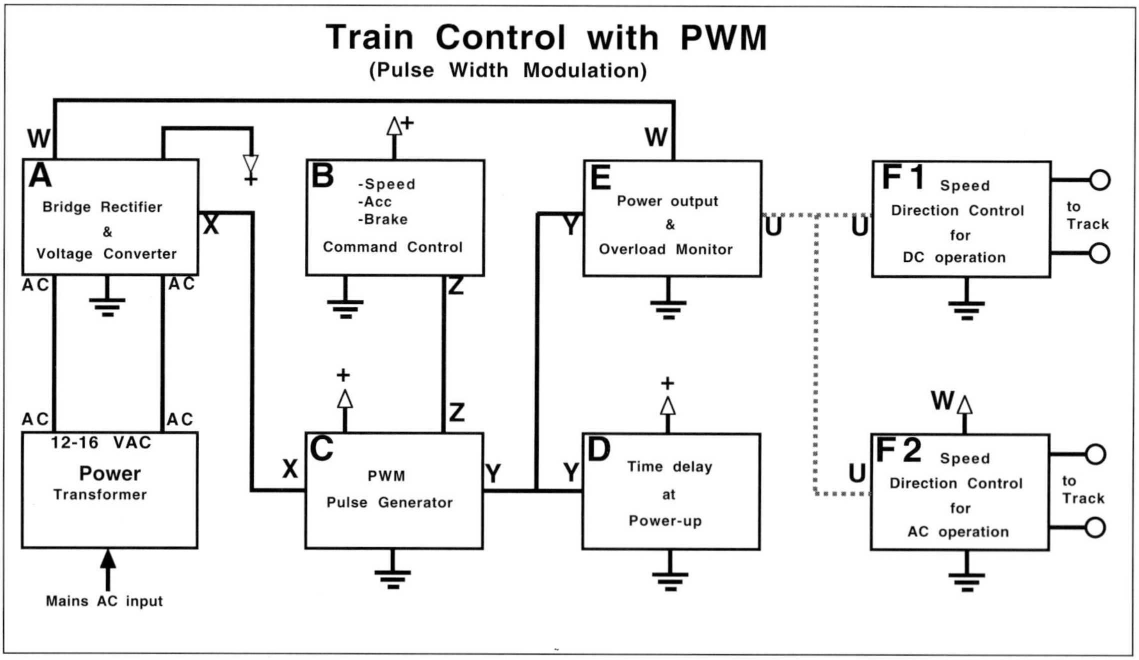

Module B

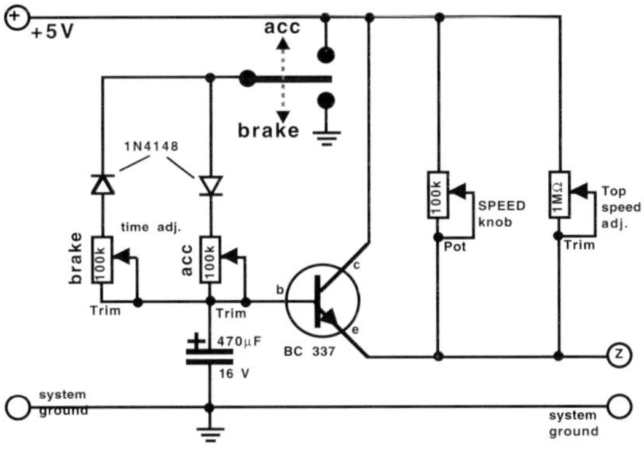

Module C

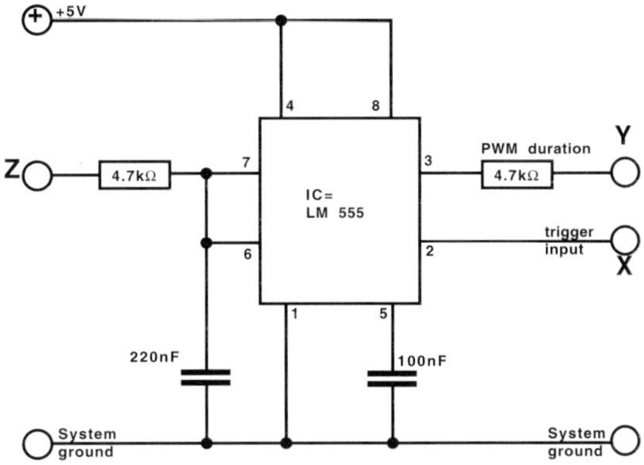

Module D

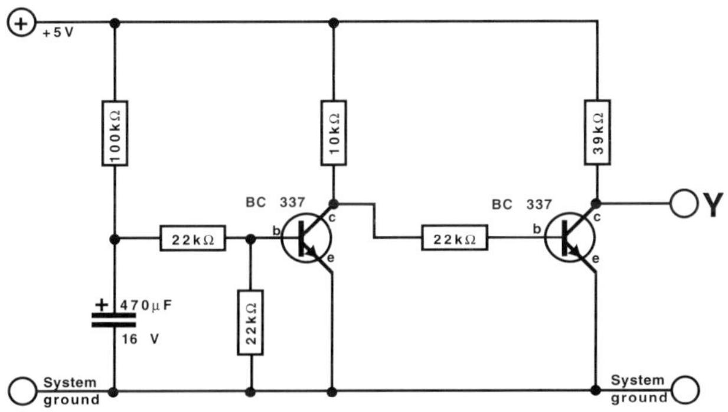

Module F1

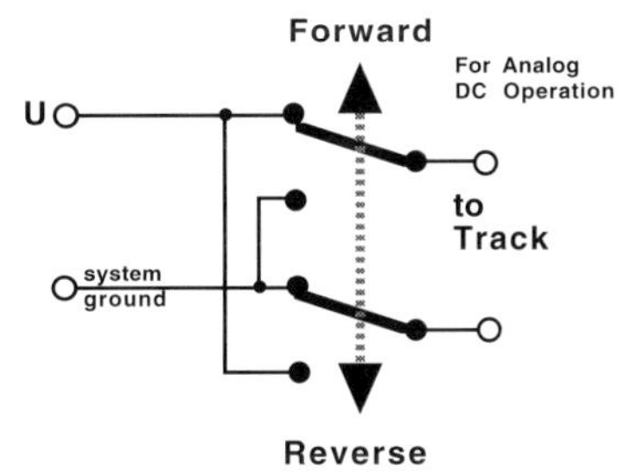

Module F2

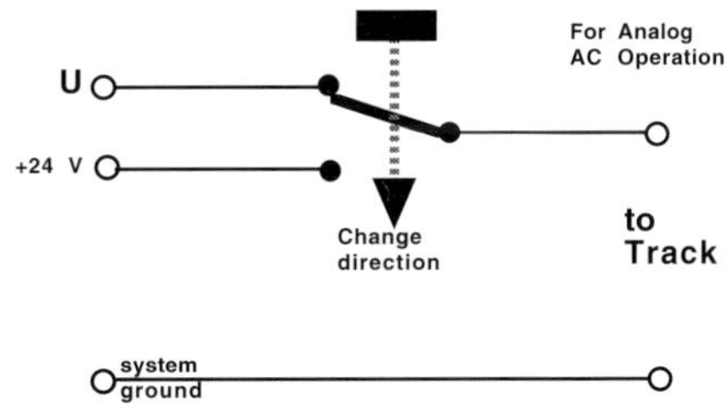

Module E

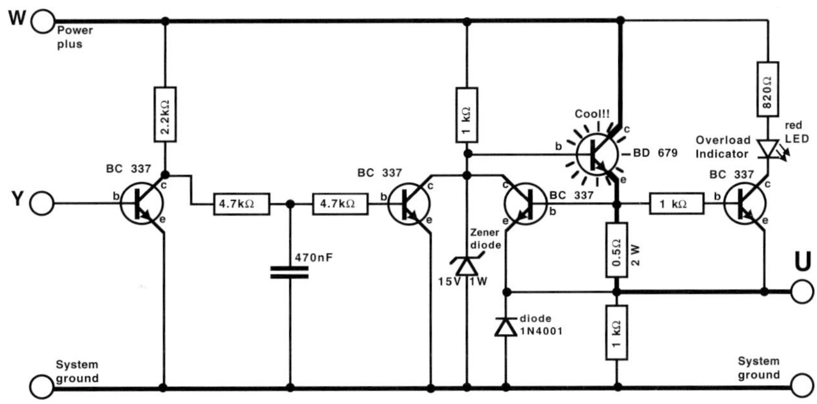

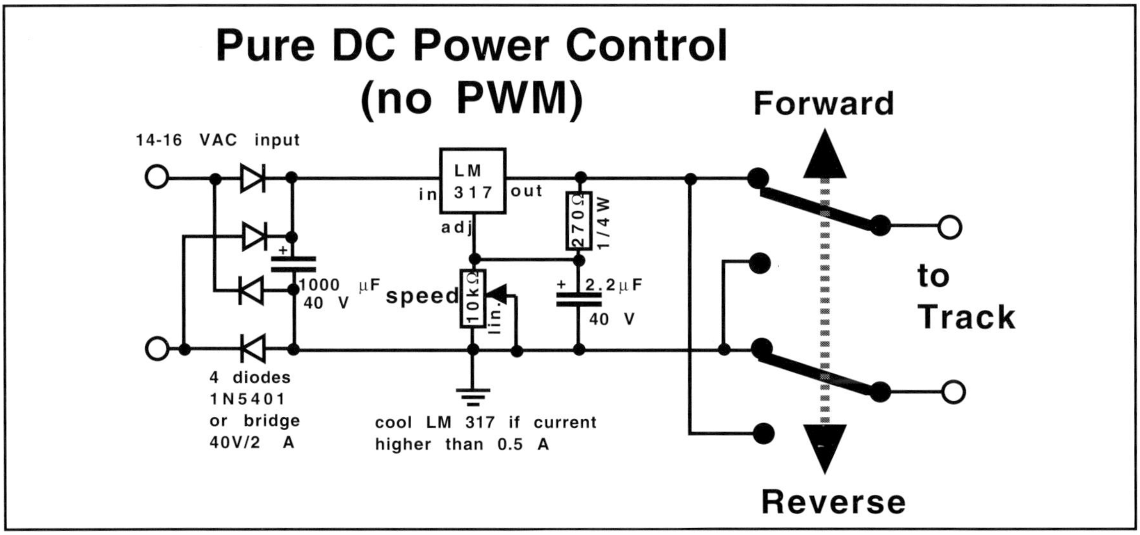

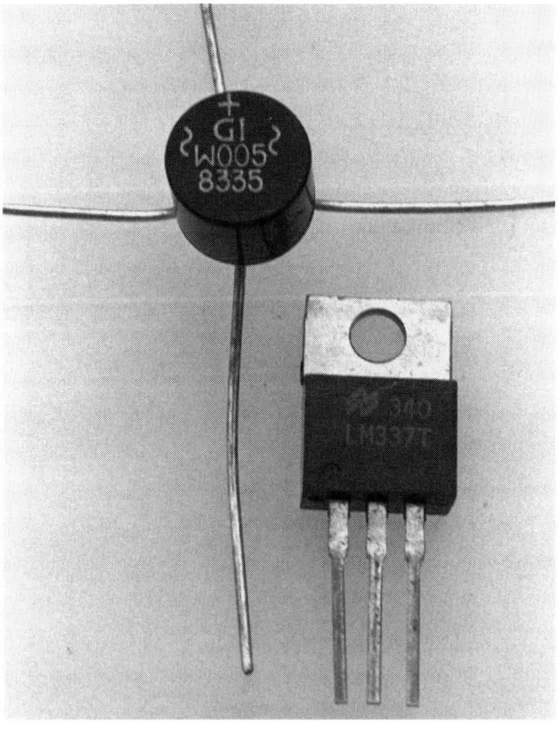

The elschematics indicates very few components versus the task this module can perform. Main components are the bridge rectifier and the voltage regulator. The latter should be cooled if the load is larger than 250 mA. As shown in the picture with components explanations you must be observant on how the LM 317 is connected since the pinning is not very logical.

As shown on the photo the rear side of the IC is a metal plate, to be connected to any massive cooling bracket. It might be handy to know that this metal plate is the same potential as the mid pin of the IC.

MINI-TIP

Basic Cab Control – none PWM output

Some locomotive motors can be damaged by pulse controllers. There are pulse controllers in this book for those of you desiring this type of controller.

When should pulsed controllers be used? Electric motors have a friction threshold that makes a relatively high starting voltage a necessity. Immediately after the rotor begins to spin, voltage can be reduced. Otherwise, the rotor will turn too fast. The problem is that it is impossible for us to manually reduce the voltage quick enough to avoid jackrabbit starts. The engine's ability to start and stop smoothly is what determines its running qualities.

Pulsed control units send out short pulses at full voltage. Motor regulation is not by voltage level but rather by how often these pulses are sent and their relationship to the no-voltage state per cycle. Controllers exist that combine both methods of control. Under certain conditions, locomotives running on pulsed power "sing".

Pulsed control is beneficial to engine running quality but potentially damaging to motors. This is because pulsing causes motors to run warmly. Most motors used in HO and N scale can handle this. Generally, one should be cautious with small Z scale motors and can motors. Seek advice if you are unsure of the motor types used in your own engines. Often, the locomotive manufacturer provides information on their products' compatibility with pulsed power.

A Mini-Project is provided for motors sensitive to pulsed power. It's built around an LM317 voltage stabilizer with a maximum current regulation of 1.5 A. Output over 0.25 A is cooled with a heat sink. This is a safe circuit - it's short circuit protected and has an internal thermal cut-out that shuts down power when heat is excessive.

MINI-TIP

Alternative wiring – paint-on leads

There are times when there isn't room for even the thinnest of leads. The photograph shows a plastic N scale power pole. A lamp was installed at the top. In this example, drilling lead access holes through the poles was out of the question.

The solution was to paint the leads in place. The German model supply firm Busch has developed a silver lacquer that suits our needs perfectly. The item number is 5900.

The lacquer begins to dry after approximately two minutes and after two hours, has a surface resistance of 0.02-0.1 ohms per square centimeter. After 24 hours, it has a hard and workable surface that can withstand temperatures of up to 300° C (570° F). Included is a bottle of thinning and cleaning solution. The silver content is approximately three grams (a tenth of an ounce) and the kit's selling price is about $15.00.

One of the positive features of the lacquer is that it can be covered with common hobby paints, making wiring completely invisible. Other thinkable applications are painting insulated railroad car wheels, adding lighting to automobiles, figures and possibly, even to bicycle head and tail lights. In passing, the lamp shown in the photograph measures 1.5 mm by 2.5 mm (0.06" x 0.10") and is meant for a 1.5 volt power supply. With a 1 k ohm series resistor, the lamp can be connected to 14 volts.

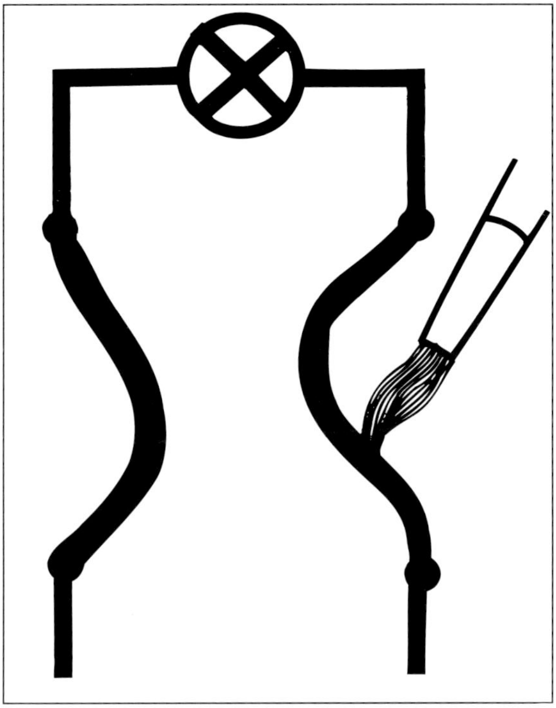

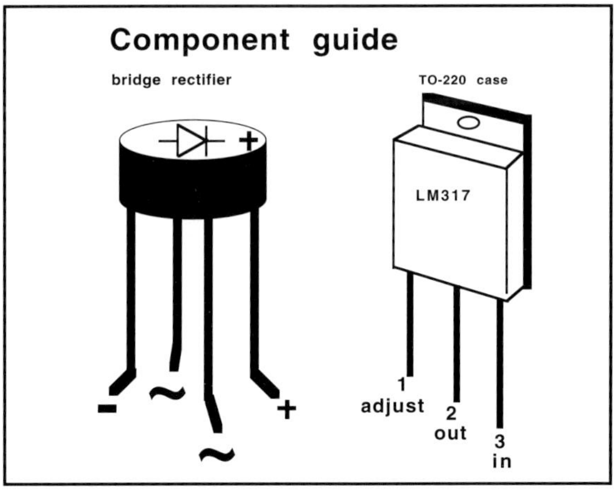

The LM317 lead configuration is uncommon enough to warrant a separate figure. The circuit can handle voltages of up to 40 VDV. This corresponds to an AC voltage of the AC voltage times the square root of 2. That means that 12-18 VAC is not a problem. Voltage (and locomotive speed) is regulated by trimming potentiometer P. Resistors are of the common type.

Block control with soft acceleration and braking

If you want to run several trains simultaneously and after one another on the same track, some form of regulation is required to avoid collisions.

This module controls trains on a block-sectioned track loop. The minimum block total is three – for two trains – but there is no upper limit. Block total must be at least one block higher than the train total.

If a train begins to catch up with another, it will come to a gradual stop and wait until the next block is unoccupied. When the block is clear, it will accelerate slowly. Acceleration and braking can be adjusted individually for each block section. A hand controller (potentiometer) at each block is featured for manual running and to also permit you to set maximum block speeds.

The block control will work on both AC and DC layouts. In its simplest configuration, all trains run in the same direction with an unbroken minus rail. Wiring is easy in that just the plus lead is run to each block. Consequently, no relays or contacts are needed.

It is advantageous to build several modules on the same circuit board and use IC 7812 to run several circuits.

A and B are control ports connected to adjoining block modules. If the track branches off at a switch, a block's A output can control several other blocks' B inputs. With diodes, a B input can even be controlled by several A outputs.

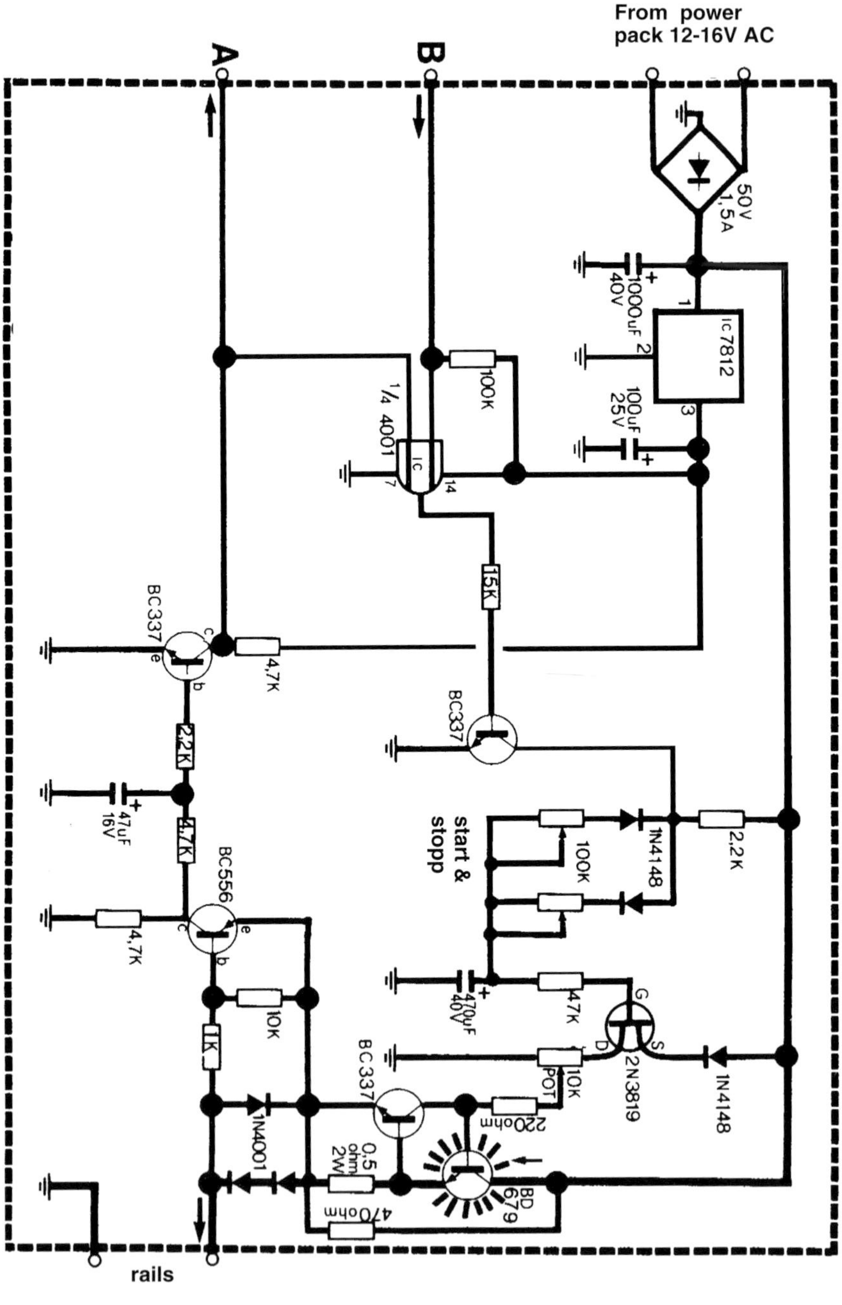

Block control K

Fires with electronics

Buildings on fire are serious matters in reality. A fire on a model railroad is just as attention getting. There are plastic kits of partially burned down buildings if you don't want to scratch build one yourself. Fire fighting vehicles are also available in almost all scales. In this project, we'll be electronic pyromaniacs.

The difficulty in modeling fires is in reproducing the flames without the usual repetitive blinking circuits. I've seen household current, candle-like bulbs used to imitate fire, but it doesn't feel right to use high voltage on a layout.

This suggested solution runs on approximately 12 volts and is built around four 555 timer circuits. Four lamps are placed in a building, partially obscured, along with a smoke unit. A dim yellow and a dim red lamp supply the basic glow. The other two lamps "flame" irregularly between two light levels. Combined with the smoke, the overall effect is quite realistic. Bits of cotton higher up are also effective.

All four timer modules are constructed in the same way but component values vary. Try experimenting with other component values. You can, of course, increase the fire intensity by using even more modules. Fortunately, this fire is easier to extinguish than others – just cut the power!

More about the fire brigade on page 58.

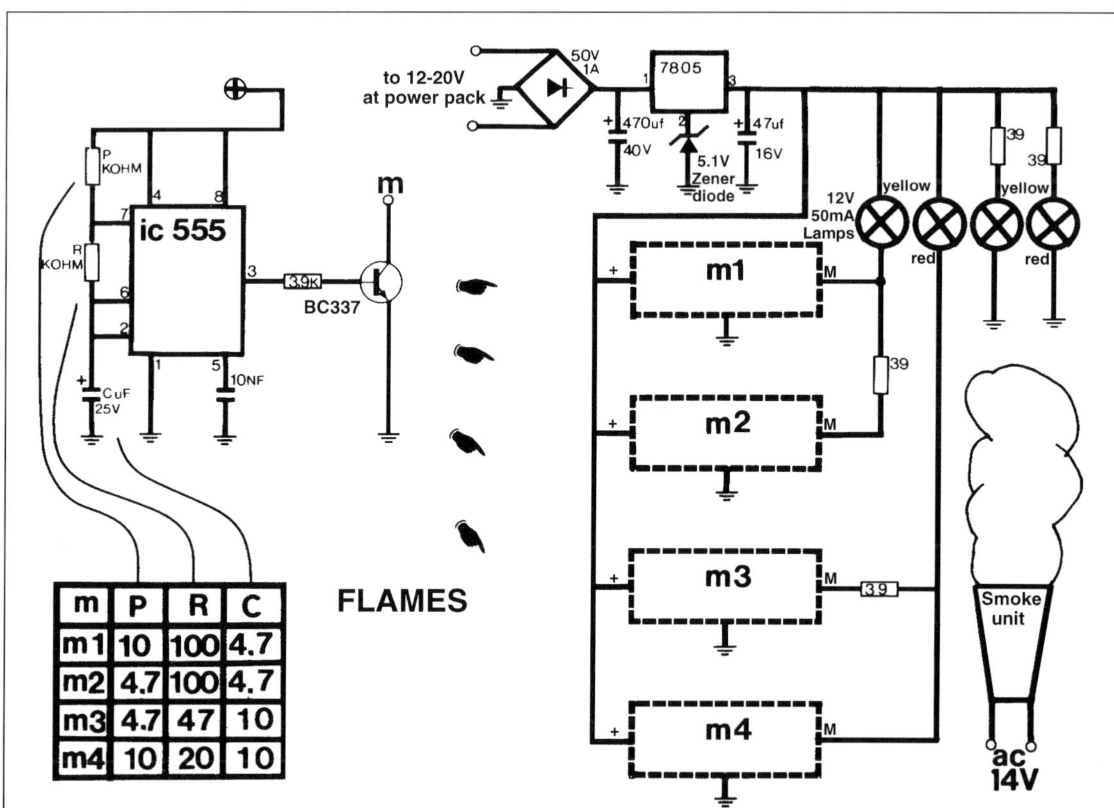

m	P	R	C
m1	10	100	4.7
m2	4.7	100	4.7
m3	4.7	47	10
m4	10	20	10

Jack up the voltage!

This voltage ladder is a practical circuit if a voltage considerably higher than that available at the transformer is required. Let's say that we need 30 volts (DC) as a reference voltage but have a transformer with just 5 volts. We would build a voltage ladder up to the sixth rung (5 x 6 = 30).

Here are a few examples of modules that I have built myself that required voltage ladders. For a fire simulation I decided to use a household current bulb at a little less than half its normal voltage. The effect was just as desired. All capacitors were 100 nF/250 volts and the diodes 1N4005.

On another construction, I needed a small lightning flash complete with thunder. The outputs were connected to two pins with the tips pointed towards each other, about 1/32" apart. The ladder was extended and the capacitors exchanged for the 400 volt variety. The diodes were 1N4007s. For safety's sake a resistor R at 5.6 ohms (1000 volts) was used to limit current in the secondary circuit.

Because the ladder can be built as high as desired (until insulation problems occur) and dangerously high voltages can be attained, I would like to warn you of the very real, potential dangers associated with high voltages. Note that only the secondary side of an approved transformer should be used.

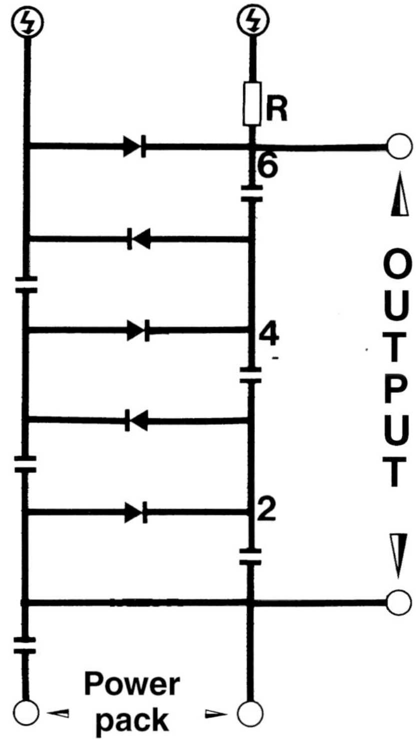

Voltage ladder

MINI-TIP

A DC ENGINE FOR AC OPERATION

Electronic Reverser Switch

In the first book of the series, I described the alterations necessary for converting AC Märklin locomotives from AC to DC operation. Now we'll do just the opposite and build a little reverser switch for DC locomotives. These locomotives have just one field winding. To switch engine direction, current through this field winding must change directions. This can be done with a double-pole, bistable switch relay but at a high cost. It can also be done with a few power transistors at a considerably lower price.

THREE-RAIL OPERATION
DC locomotives normally (Trix is an exception) use the running rails for current pick-up. This is why the wheel axles are insulated. With Märklin's three-rail system, the running rails have the same electrical potential and are connected with one another. In the DC locomotive, we'll do the same and connect both leads to the same AC connection point at bridge D1. Now for the hard part. A collector shoe must be mounted under the engine for connection to the second AC bridge connection point. Take a look under a Märklin locomotive for ideas on how to go about mounting the shoe. I would suggest buying a Märklin, sprung collector shoe.

FLIP FLOP BASED
We'll use an IC of the CMOS type that controls running direction in accordance with the "every other time" principle. Pins 1 and 2 are complementary outputs. When one is high, the other is low. They switch potential each time a pulse is received at pin 3. If pin 1 is high, transistor T5conducts. In this case, it conducts to ground because the emitter (marked with an arrow) is connected to ground. The four power transistors (they don't need heat sinks, by the way) work in pairs. In our demonstration, T1 conducts at the same time as T6 and T4. We can now follow the current from plus via T1, through the motor M, and down to ground through T4. After the next switching operation, T2 and T3 conduct instead, and current passes through the motor in the opposite direction. In this way, running direction is reversed.

SOUNDLESS REVERSING
You may have seen Märklin locomotives previously and noted that they often make jerky and noisy direction changes. This isn't the case with this reverser. Transistors T7 and T8 ground the base connections for both T5 and T6 simultaneously when the reversing signal is received from

Engine Onboard Control Module for AC operation

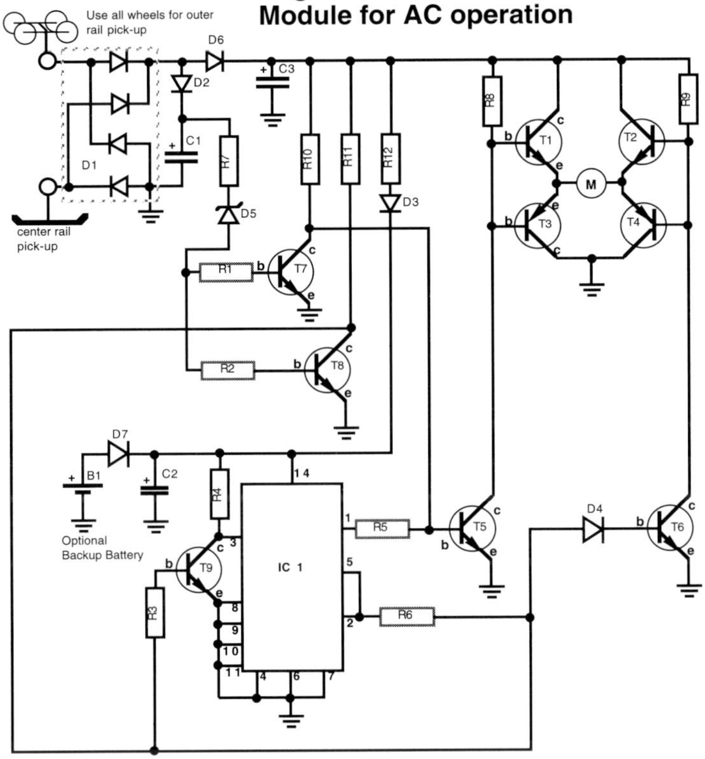

Converting
DC engine to AC

Components list

R1-R6	**100kΩ resistor**
R7-R9	**4.7kΩ resistor**
R10-11	**47 kΩ resistor**
R12	**1kΩ resistor**
D1	**1.5 A /40 V bridge**
D2	**1N4001 diode**
D3-D4, D7	**1N4148 diode**
D5	**27 V / 400 mA Zener diode**
D6	**If M < 1 A then 1N4001 diode**
D6	**If M > 1 A then 1N5401 diode**
T1, T2	**BD679 transistor**
T3, T4	**BD680 transistor**
T5-T9	**BC337 transistor**
C1,C3	**10µF / 40 V tantal capacitor**
C2	**0.68 µF/ 35 V tantal capacitor**
IC 1	**Cmos IC 4013 (1/2)**
B1	**2.4 V button cell**

the transformer. This blocks current supply to the motor. At this instant, IC1's full potential is over either R5 or R6 (which of the two is dependent upon the "every other time" principle). The circuit is designed for this, enabling it to also switch to the other position with a pulse at pin 3 from T9. When the controller is released from the reversing position, functions return to "normal". So that IC1 can remember its prior control value while running current is turned off when changing direction, a back-up voltage supply is needed. Two miniature batteries B1 (2 x 1.2 volts = 2.4 volts) are employed. These will last a long time. IC 4013 in its resting state draws current corresponding to the batteries' leak current. You can read more on this in a mini-tip sidebar further on. When the locomotive is running, "normal" current is led through R12 and D3 to IC1 and T9. If a shorter control value memory is acceptable, greatly increase the value of C2 and eliminate B1. You can save battery current by equipping B1 with a miniature switch for use during lulls in operation.

You will most likely find that the to AC modified engines will run very slow and smooth in the idle speed register. Older Märklin engines on the same layout might take off directly at somewhat higher pace.

If the rotor is of a kind like the 3-pole one on this photo you can improve low speed performance by purchasing 5-pole rotor, originally made for the engine digital kit 6090, but also very useful for analog operation.

Märklin two-train operation with one power pack

This one is for those of you that are new to the hobby and have yet to amass a large collection of locomotives and power packs. With this little switch, you can operate two trains and run them from the same power pack.

The material list is limited to eight diodes (1N4001) and a two-pole switch. A few simple tools and a fine tipped soldering pencil are also required.

We'll start the project with the engines. You may want to begin on an older one in case you run into problems. Look at the illustration showing the locomotive wiring diagram. Remove the shell and locate the relay and its contacts to the field windings. Solder two diodes (in opposite directions) to the SAME RELAY CONTACT. Leave the other relay contact vacant as shown in the lower diagram. Solder the other ends to the field windings. Repeat the process on the second locomotive. You'll use four diodes for two locomotives.

The remaining five components will be connected between the power pack and the track as shown in the second illustration. A rectifier bridge may be substituted for the four diodes. The AC connections are connected to the power pack's O and B outputs. Finally, a switch is connected between the rectifier bridge and the track. This switch reverses train direction.

You can use the controller lever as usual to vary speed. At the lever's far left, spring back position, one normally switches train direction of travel. This position now switches running current between the two locomotives.

o begin operation correctly, place the first locomotive on the track. Use the far left controller position to set it in the OFF state. Remove it and place the second locomotive on the track. Set it in the ON state. Test the reverser switch. You can now place the first engine back on the track. They can now run over the same rails and carry out interesting switching maneuvers, for example, using just one power pack

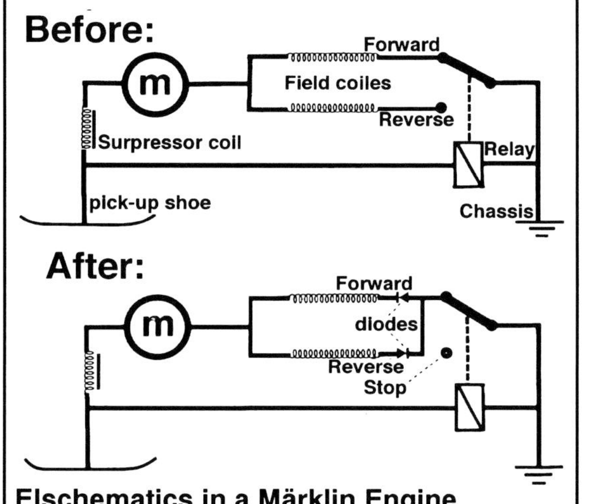

Elschematics in a Märklin Engine

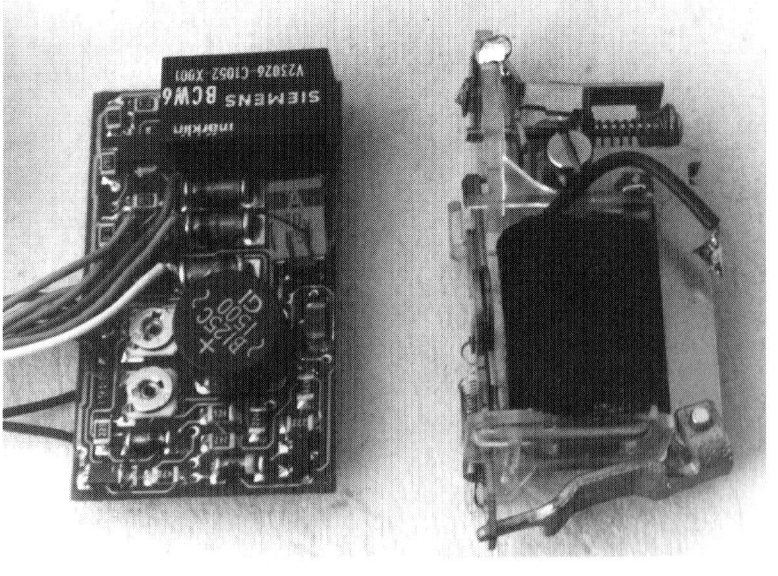

The relay (right) is the classical type found in most Märklin engines. In this project it gets a slightly different task selecting between "run" and "stand-by" instead of Forward-Reverse control.

The engine might be equipped with an electronic speed direction selector (left), but basically there is no difference how it is employed for use in our project. Cut one of the two coil leads free and use just one output of the module to distribute the power via the 1N4001 diodes to the motor field coils.

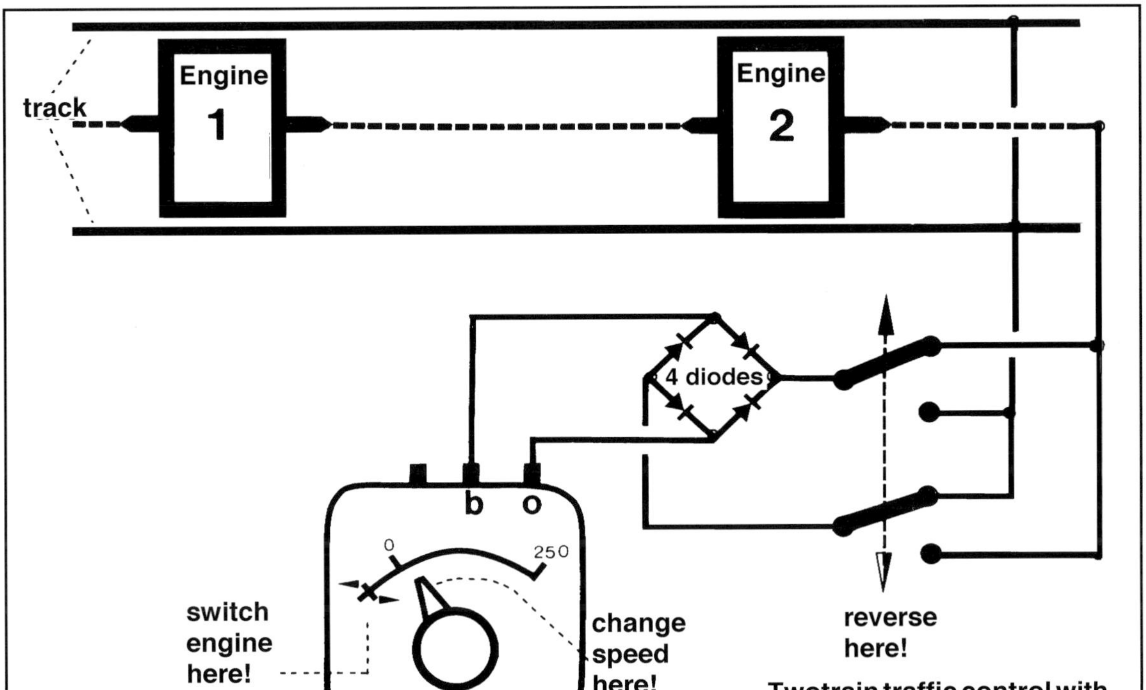

Twotrain traffic control with one transformer

MINI-TIP

The thyristor – how to test it and how to replace it

Thyristors are used in the book's construction projects and it can be appropriate to give them a closer look. For hobby applications, they are rather uncommon.

If you suspect a faulty thyristor, this simple test circuit can be of help.

Connect the thyristor as shown. Open B and ground A (to minus). The first test is close B and check that the lamp remains off. If the lamp is functional, it should light when switch A is moved to the plus position. The lamp should remain on when A is switched back to ground.

If the thyristor is faulty or if you are lacking one for one of the book's projects, a replacement can be "manufactured" with two transistors and a diode as shown. The figure shows a basic circuit using replacement components. If you are replacing a considerably larger power thyristor, the transistor equivalents must be increased to match the required effect.

Test Equipment

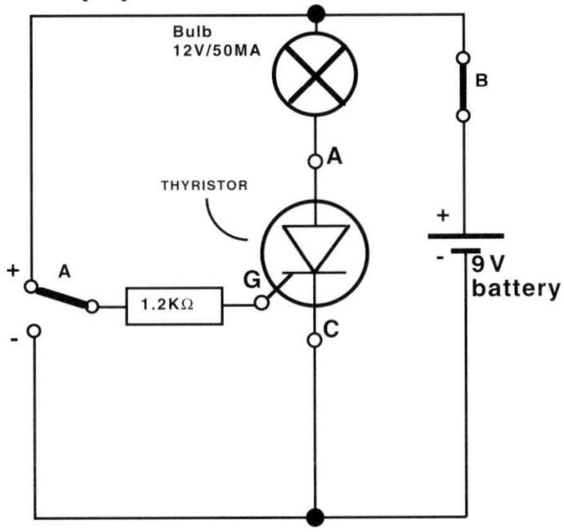

How to substitute a thyristor

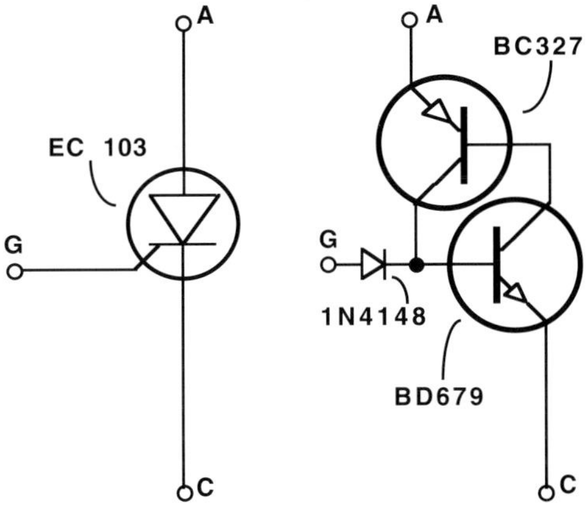

Let the trains control platform lighting

The basic idea is that platform lighting at the arrival track is switched on as a train approaches the station. The classic solution involves mechanical detectors and relays. Here is a circuit that takes advantage of modern electronics, is inexpensive, but most important, it works silently – no relay clicking!

The circuit is suitable for all train systems and scales, and can even be completely separated from the rest of the layout electrical system. If you are so inclined, you can even power it from a 12 volt battery and eliminate the need for the stabilizer IC 7805.

The circuit will work regardless of the direction of travel of the approaching train. It is controlled by two reed switches at each end of the platform. These are attached close to the rails. Under the locomotives that control lighting, small magnet are mounted. Locomotives that pull passenger trains would be equipped with magnets while freight engines, that normally just pass through the station, would not.

The IC is connected to an RS flip-flop - a bistable relay - and reacts when trains are arriving at and departing from the platform. The transistor functions as a switch and can handle up to six lamps. These should be rated for the same voltage as the voltage connected at the U inputs. A flasher module can also be connected at output Y for operating, for example, a flashing warning light for passengers crossing the tracks between the platforms.

The circuit can also be used for a number of other functions. Grade crossing gates can be lowered and signals switched on via output Y, and the gates raised and the signals switched off via output Z. The circuit could also be used to control a drawbridge.

At start-up, the RS flip-flop can take the "wrong" position. For this reason, the push-button switch R is included in the circuit. With a few more components, the reset switch can be replaced with automatic circuitry. This is taken up later in the book.

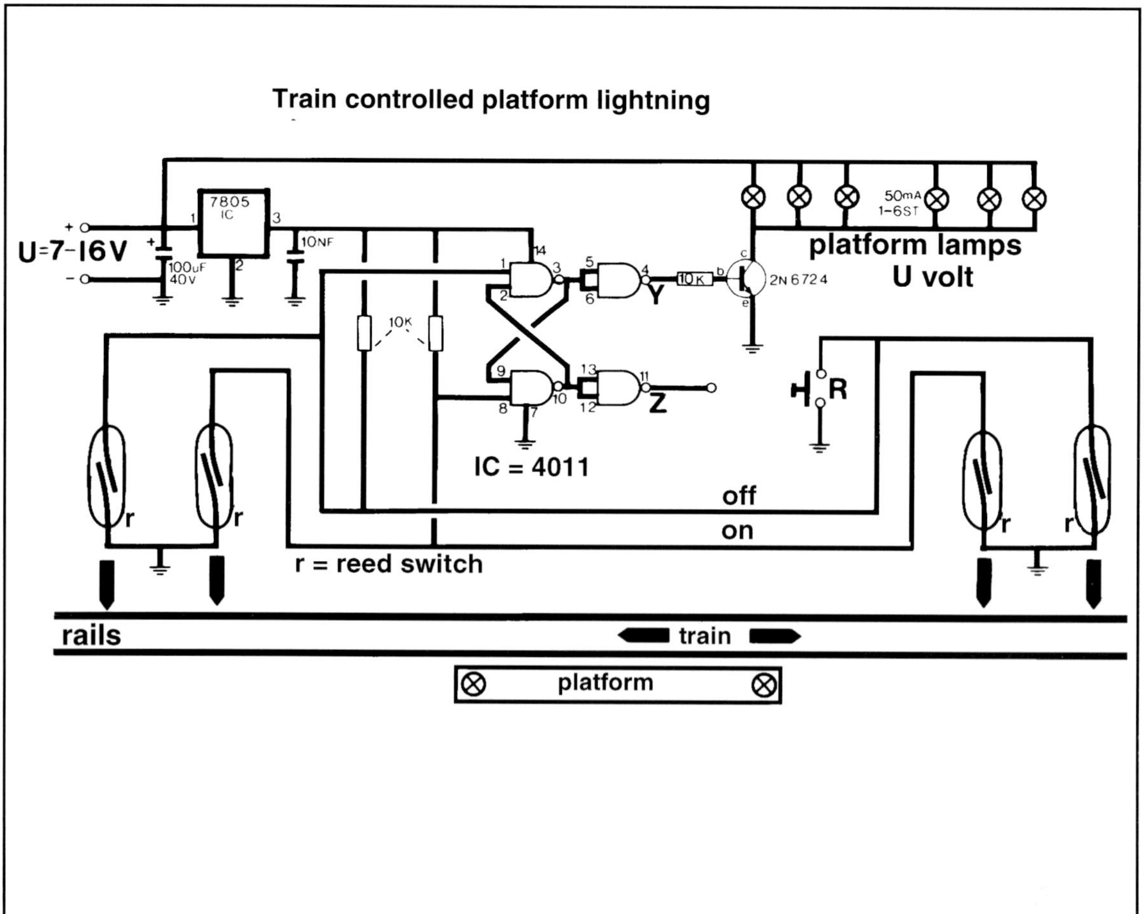

Flashing grade crossing signals

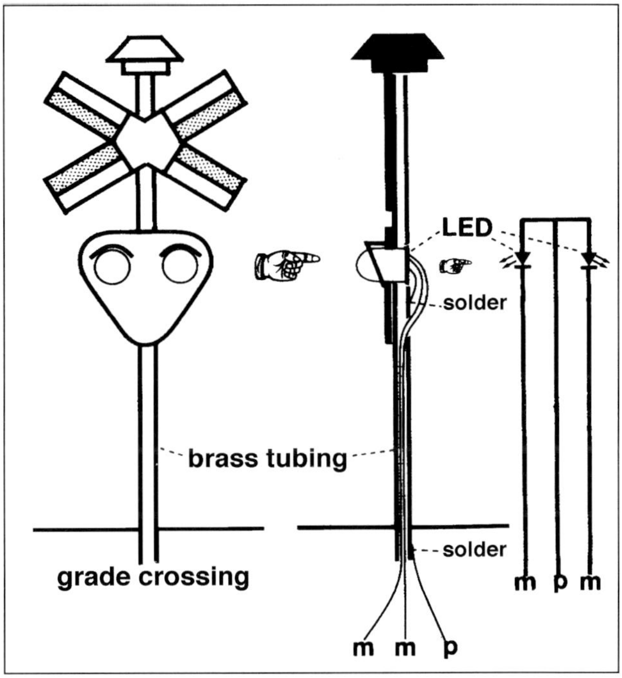

Warning signals are common features of grade crossings. In HO, there are plenty of commercial grade crossing signals to choose among. Modelers in N and Z scale are not as fortunate. The accompanying figure shows a scratchbuilt signal, based on a Swedish prototype, using a length of brass tubing for the pole. Within the pole are two thin leads. The pole itself serves as a third lead. The crossbuck can be constructed from either brass sheet or heavy paper stock. In N and Z scale, 1 mm LEDs are used and in HO, 3-5 mm LEDs.

The blink generator is built around two, free swinging transistors complemented by electrolytic capacitors and resistors. The two 100 ohm resistors limit current through the LEDs. By changing the value of the 10 kΩ resistor, the flash rate can be varied.

The module can be connected to either DC or AC power sources.

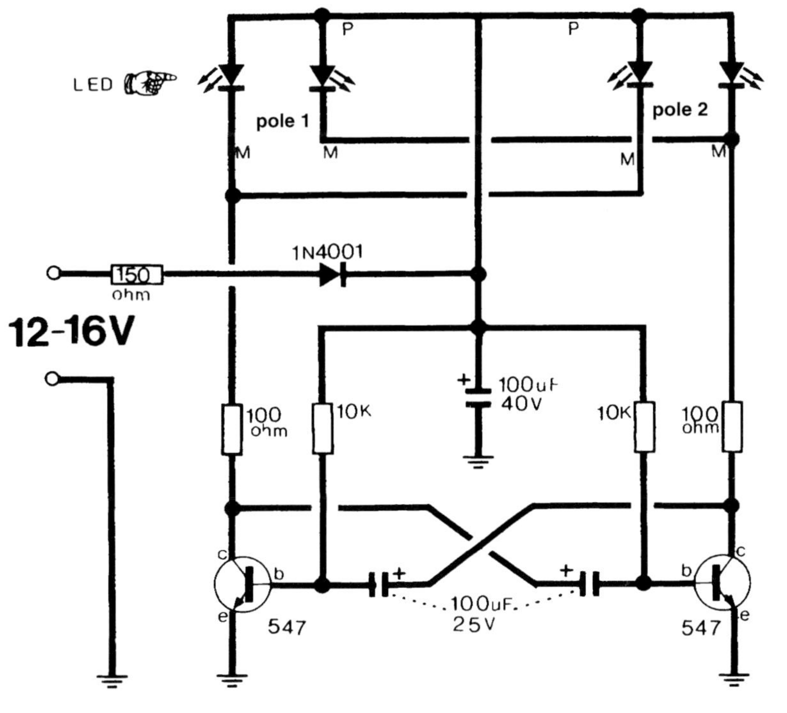

Flash module for grade crossing

A control panel overload indicator

Creeping current, overloads and short circuits are things we would like to avoid on our layouts. Trouble shooting can be complex. In the worst event, we're lead to the fault source by a wisp of smoke.

Here are two circuits that can ease trouble shooting. The upper current indicator circuit X, consists of just three components and is connected parallel with each fuse. An overload breaks the normal current flow by either melting the fuse or, in the case of thermal or electronic fuses, by disconnecting the current. As long as the short circuit remains, at least one of the LEDs will light - which one is dependent upon current direction. The value of R varies with working voltage. For 12 volts, 680-820 ohms is appropriate - the higher the working voltage, the higher the value of R.

The lower circuit works in just the opposite way. The LED lights as soon as a load is connected to the power source. The series resistance value is calculated as in the first circuit. You build the coil yourself by winding lacquered copper wire (0.8-1 mm in diameter) around a small, signal-type, reed switch. Total windings varies with desired load indication size and the magnetic force needed to trip the reed switch. The principle behind this circuit dictates that for indication of lesser current values, the number of windings is increased.

The following test results can serve as a guide. At 14 volts, I needed 60 windings for a triggering current of 100 mA and 40 windings for 200 mA. The diode can be a 1N4001.

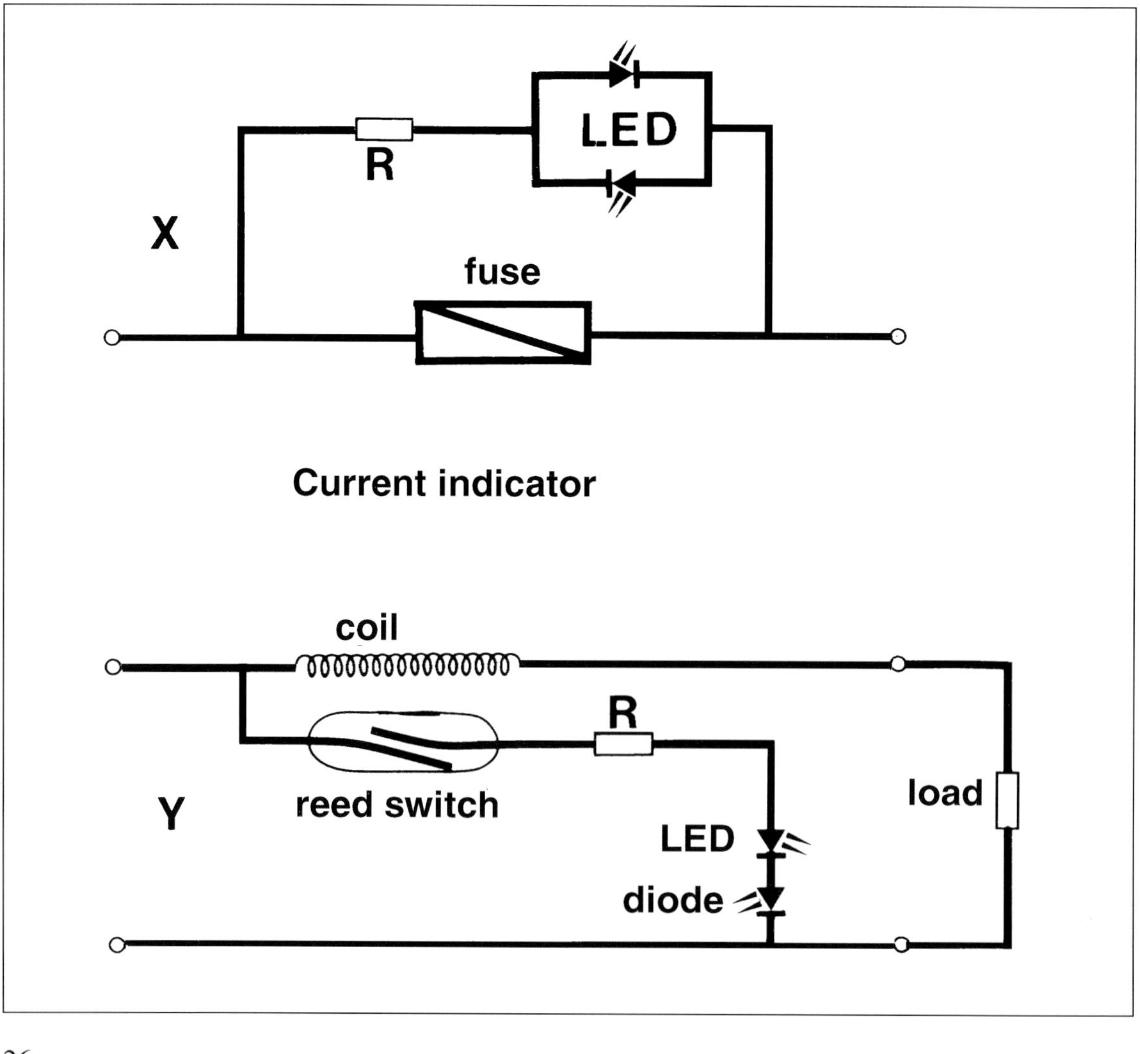

Infra-red control of grade crossing signals

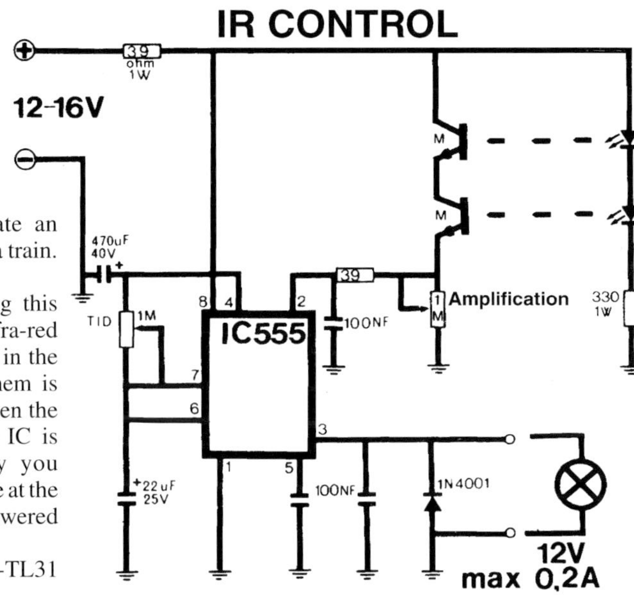

The most common way of controlling grade crossing signals and gates is probably with contact rail detectors. In some model railroad systems, only locomotives are detected by this type of detector. If cars are coupled to the front of a locomotive or if a locomotive stops just after the grade crossing, the warning system will indicate an unoccupied crossing even if the road is blocked by a train. And that doesn't look realistic.

An inexpensive and practical way of avoiding this problem is by using infra-red detection. Place an infra-red semi-conductor at each side of the track as shown in the figure. As long as the invisible beam between them is unbroken, the timer circuit is in stand-by state. When the beam is interrupted by a car or locomotive, the IC is activated for a length of time determined by you (approximately 2-30 seconds) and power is available at the output. A flasher module could, for example, be powered by this output.

Suitable components are: OP804-OP133, TIL81-TL31 and BPW34-CQY99.

The trimming potentiometer on the receiver regulates the amplification needed to activate the IC (amplification is dependent upon the distance between transmitters and receivers). Carefully make both vertical and horizontal adjustments to the infra-red transmitter so that the beam is pointed directly at the receiver. The shrinking tubing is important to screen above all, the receiver transistor from general room lighting.

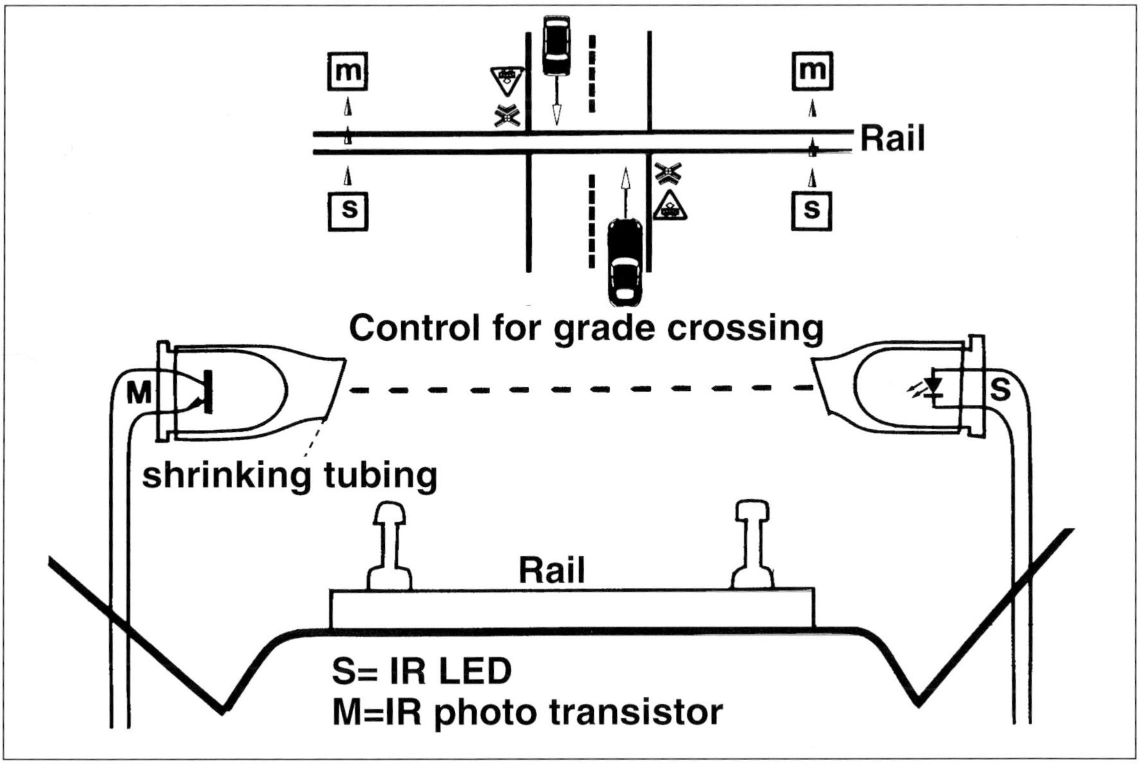

Module for Märklin Power Packs

Running DC locomotives with Märklin AC power packs

How can you use an AC controller when you run DC engines? One way is to use diodes to rectify running voltages and a switch for reversing. But the result is only a compromise.

Living in a country where used Märklin power packs are both plentiful and inexpensive, I've often wondered if there was a more elegant solution - a solution that would even permit using the Märklin unit's built-in reverser.

This is the result of my pondering. As shown in the photo, the entire module fits on a small, 2 3/3" x 3" circuit board. In the wiring diagram, I've organized the components by function for reasons of clarity. The rectifier section consists of diodes and a voltage regulator (7812). Electro magnetic accessories can be connected as desired at output M. Other letter designations are related to the controller section.

The second diagram shows regulation with two standard ICs and a handfull of discrete components. The LEDs indicate direction of travel.

The power section consists of little more than four transistors for track current regulation. These can control up to two N scale engines without cooling flanges. For running larger HO engines, equip them with flanges.

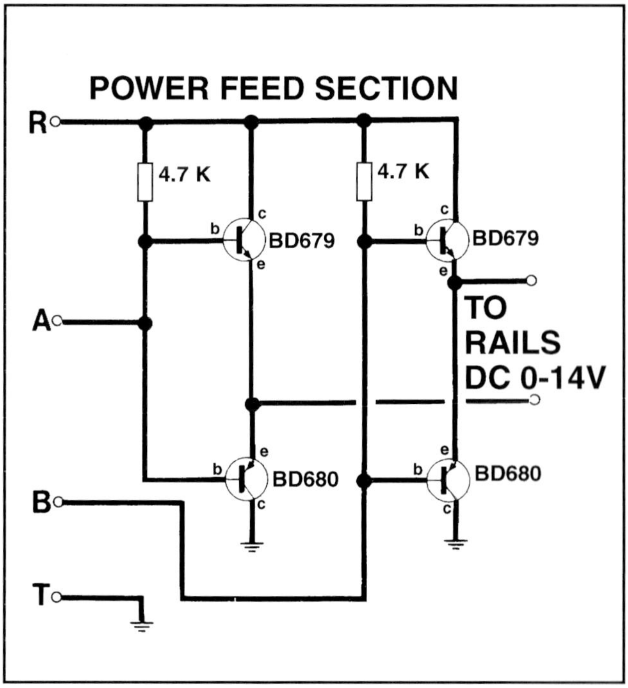

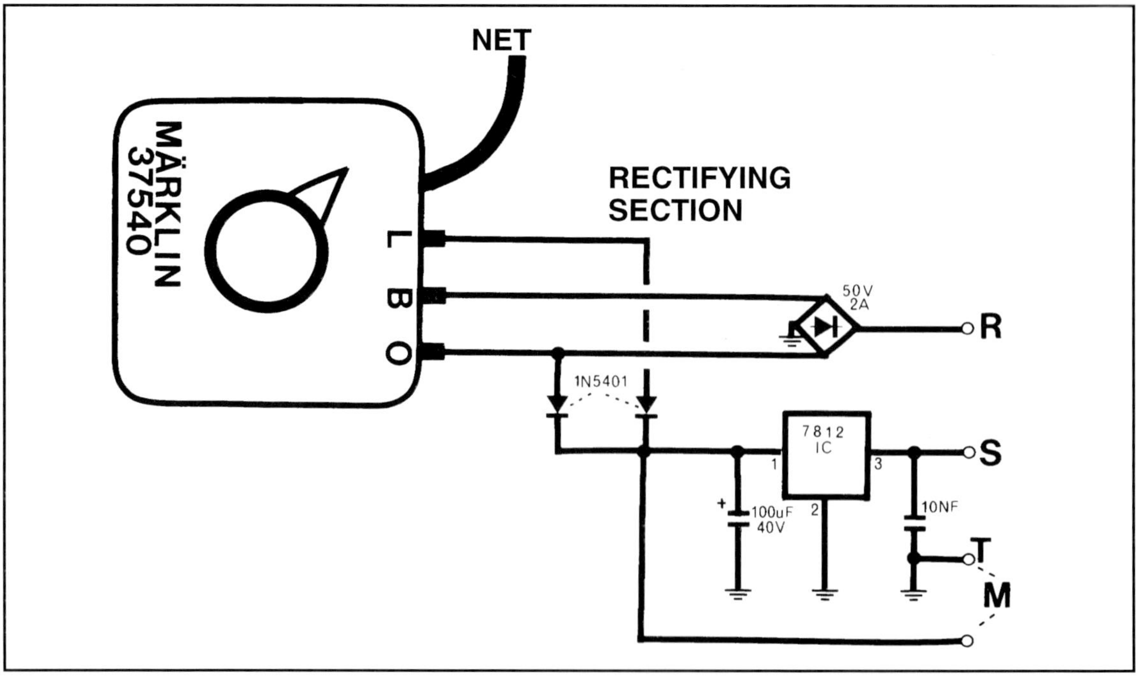

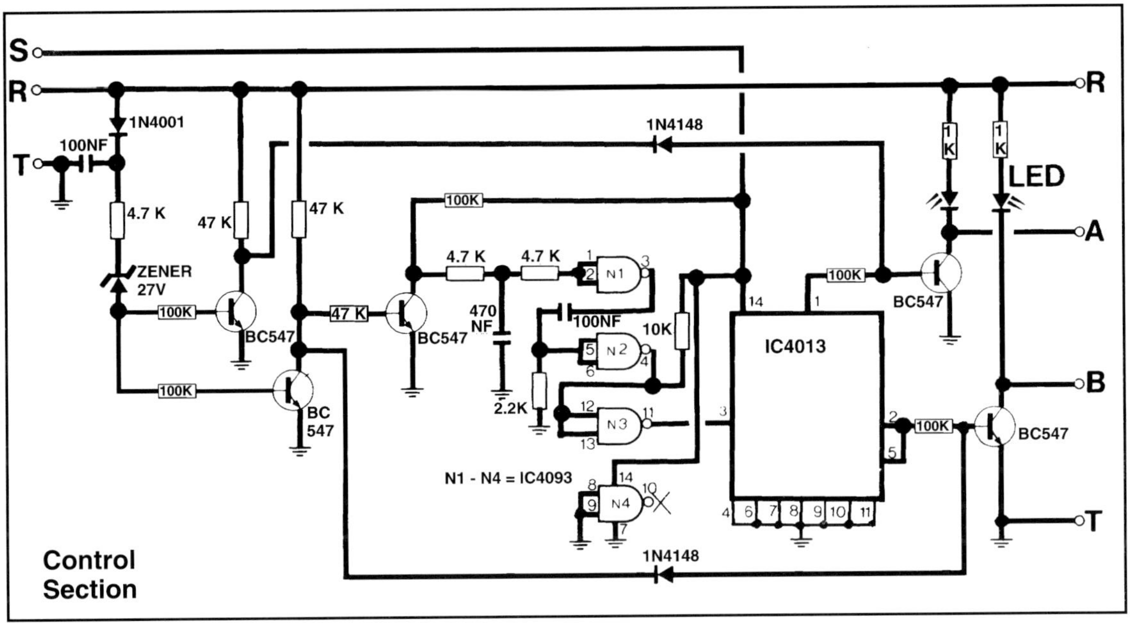

All operating functions are carried out with the controller handle. In the event of short circuits, the unit's thermal circuit breaker steps in. The power transistors can handle up to 4 amps.

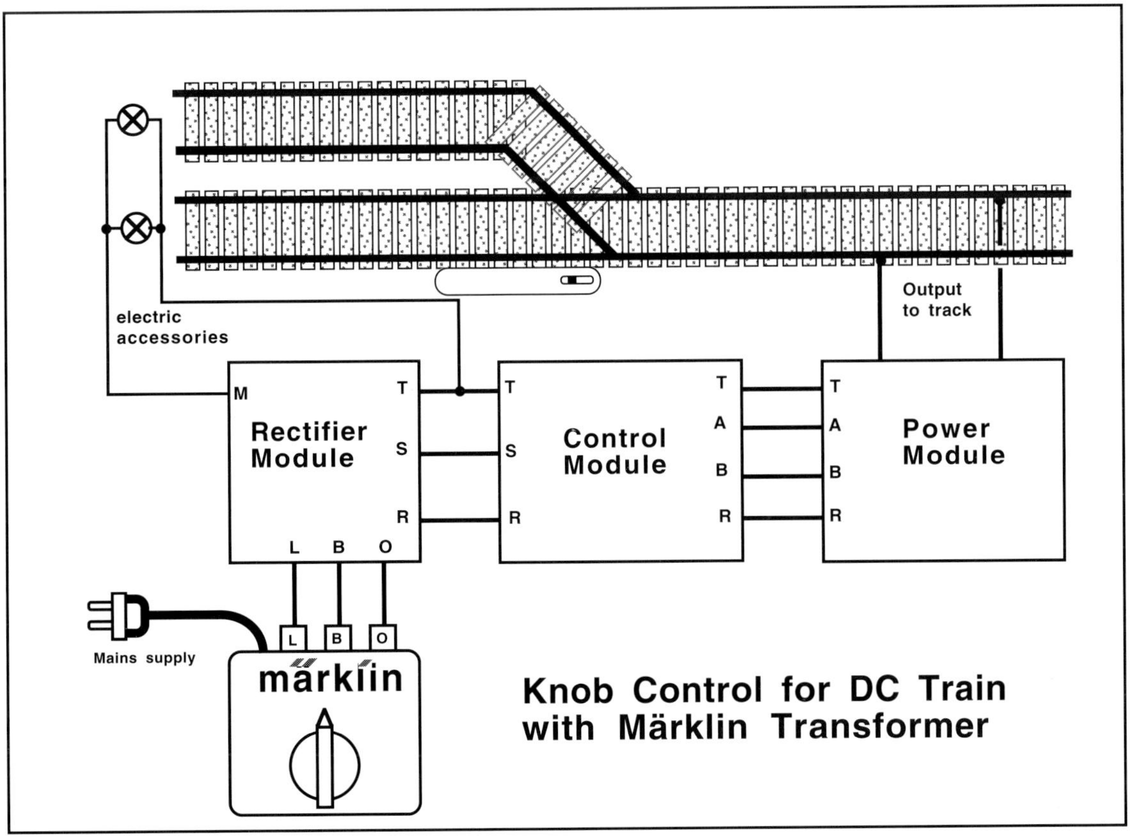

An electronic fuse with a reset feature

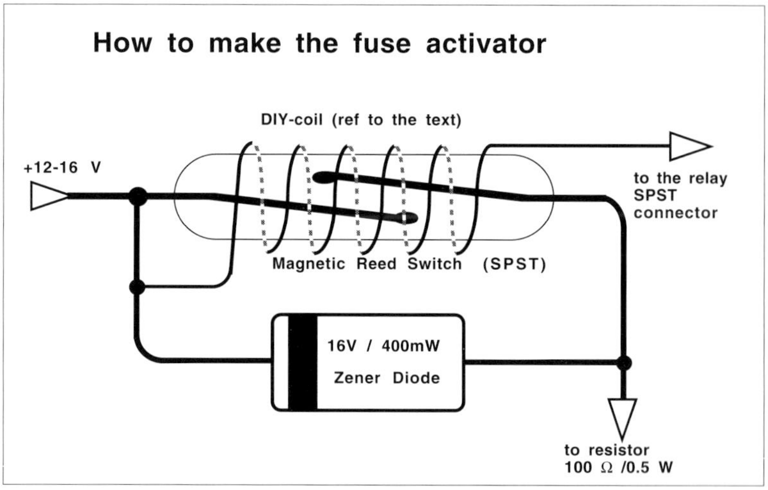

Controllers for model railroads usually incorporate a built-in thermal fuse that breaks current a few seconds after an overload condition occurs. When the bimetal in the fuse cools, current is reconnected. This offers cheap but rather crude overload protection. Modern electronic components are often unable to handle short circuit current at this duration. Long thin leads can also overheat.

I found this electronic fuse in use on a friend's layout. It has the advantage of disconnecting power as soon as an overload occurs. It also features built-in overvoltage protection. Maximum voltage is regulated by a 16 volt zener diode that triggers the relay and lights the red LED. The reset button reactivates the fuse and the green LED lights. A thyristor handles the disconnect function and is controlled by a wire-wrapped, reed switch coil. You can make this coil yourself using thin, lacquered copper wire. This fuse is suitable for use in the 0.1-5 amp range. For higher currents, reduce the number of coil windings. Start with 50 windings and make adjustments as needed.

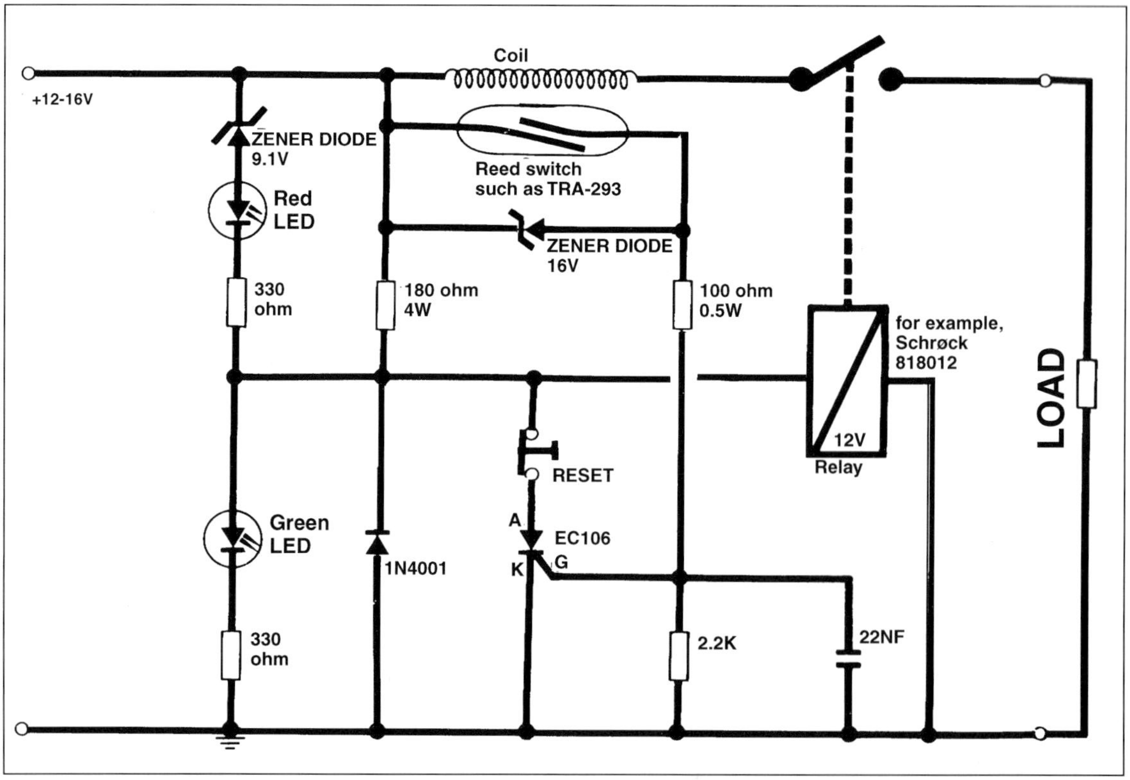

An inexpensive electronic relay for universal application

Intermediate sized model railroad layouts can require numerous relays. Relay control is dependable and trouble shooting is usually at the knowledge level of high school physics lectures. The disadvantage is cost (especially for double-throw, bistable relays) unless you are fortunate enough to come across a source of used relays.

Within certain limitations, these relays can be replaced by considerably less-expensive, electronic substitutes. This is an example of an electronic relay based upon the well-known LM555 timer. As it is used in this circuit, the output U is low (grounded) at start-up. If the control input E is momentarily or continuously activated, the "relay" will throw and output U will now take a high level (equal to the supply voltage of 12-16 V). Even if input E is disconnected, the output will remain high. The output drops to low when control input F is activated, in the same way as described for E.

What we have now is a controllable output that can be used in many different ways. A 12 volt (that's to say, supply voltage) lamp can be connected as indicated by L1. It will always be on when powering-up, can be turned off via E and can be turned on again via F. A lamp connected as L2 will be on at start-up, turned off with E and turned on with F.

Output current at U is approximately 100 mA for the CMOS 7555 and 200 mA for the standard version. This means that up to four normal, 50 mA lamps can be powered from each unit.

For larger loads, a monostable (spring back) relay can be connected as shown in the power module figure. The transistor can be eliminated for very small relays but the protector diode should be retained. We can now regulate larger loads, such as motors and motor reversers, with contacts A and B.

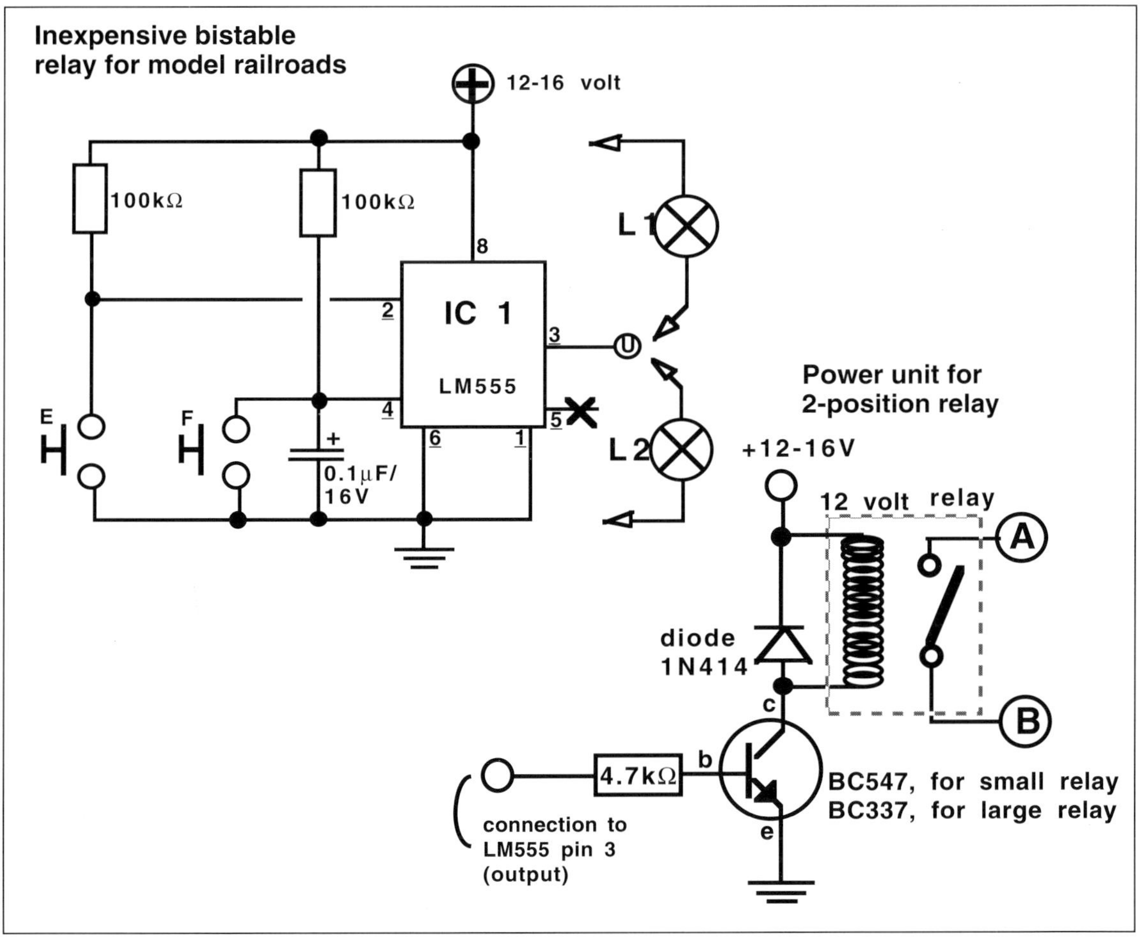

Three signal indications from one lamp

This is a rather unusual signal in that exact prototype versions are non-existent. It is a signal with a single lamp - an LED - that can provide three signal indications: red, green and yellow. When unpowered, the LED is white.

The LED is actually two LEDs - one red and one green. When you buy one of these LEDs, it is important that you get the two-lead version. Three-lead versions are available but not suitable for this circuit.

The two control inputs, R and G on the schematic, are used to produce the various signal indications. The resulting signal colors brought about by varying the signal voltage combinations at R and G are shown in the code table.

What is especially interesting is the effect achieved when both R and G are at high signal levels. The LED shines yellow - a combination of red and green. The signal inputs must have a constant, active signal level to produce the desired color. This signal can come via a relay or a bistable, electronic switch such as that shown in the schematic. This switch is activated by either a contact rail, infra-red or reed switch/magnet detector. The system can be easily cascade wired, meaning that several signals can be wired7566 together, one after the other, and effect each other. With the bistable switch, the signal system can also control train movement by letting the outputs trigger a power switching, block feed relay. For block control, use a relay control voltage that matches the system voltage.

Component wise, this project is modest, simple to build and easy to trouble shoot. We use the most common of ICs, the 555 timer, in duplicate. Initially, both of the IC pin 3 outputs are grounded (low level state). Both R and G are unconnected and voltage over the twin LEDs and series resistor is zero. When G is set to the high level state, the LED shines green, and red when only R is high. When both inputs are high, the oscillator switches current direction so quickly that the alternating green and red signal indications are conceived as yellow.

3-color railroad signal with one LED

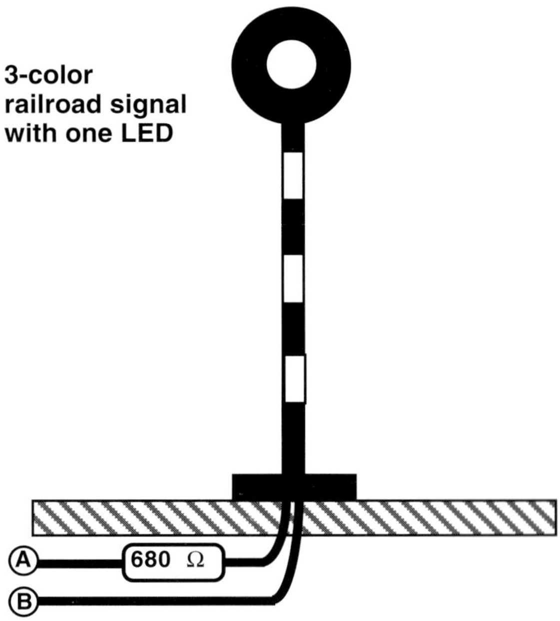

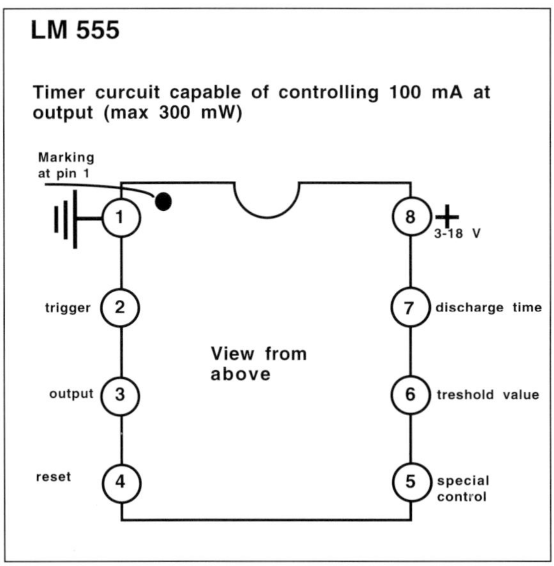

Code table for 3-color railroad signals

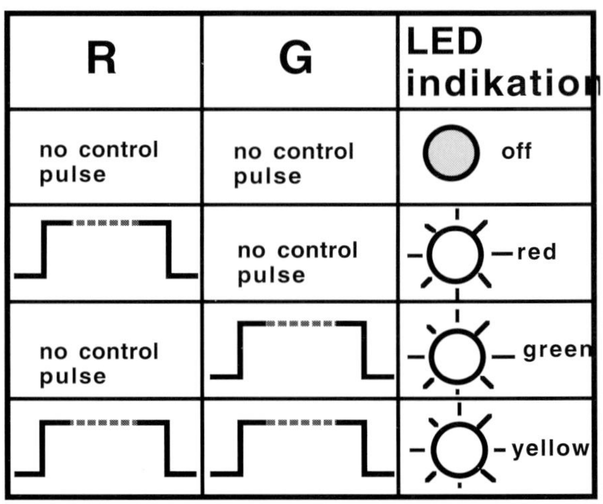

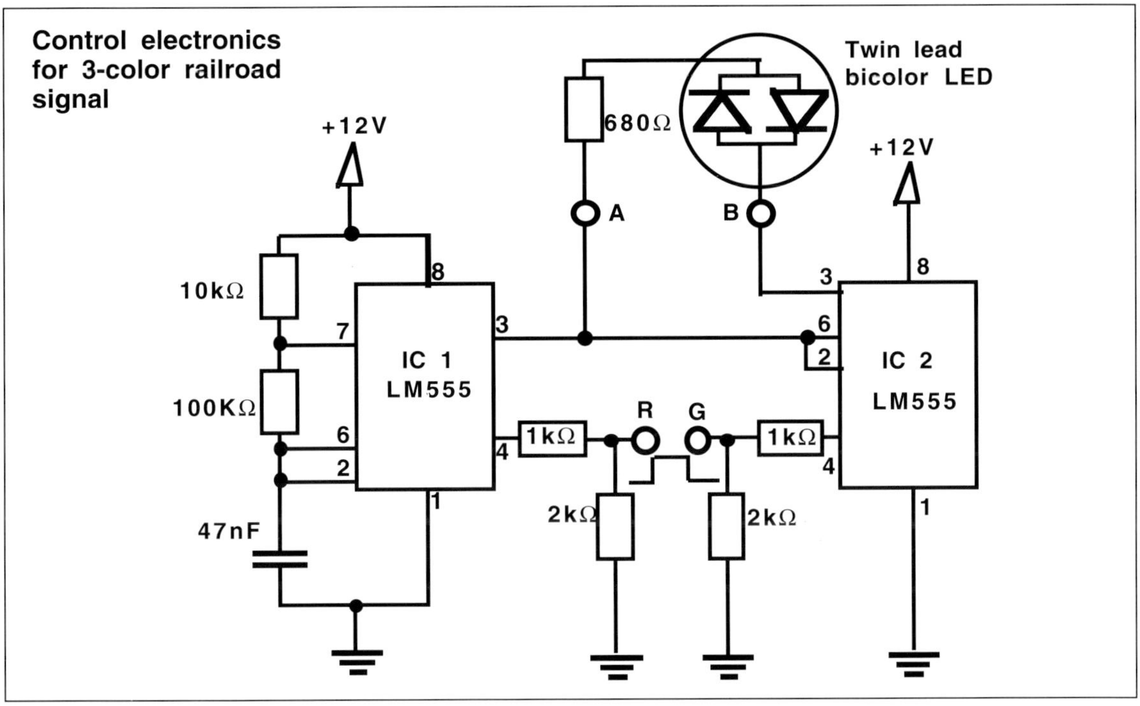

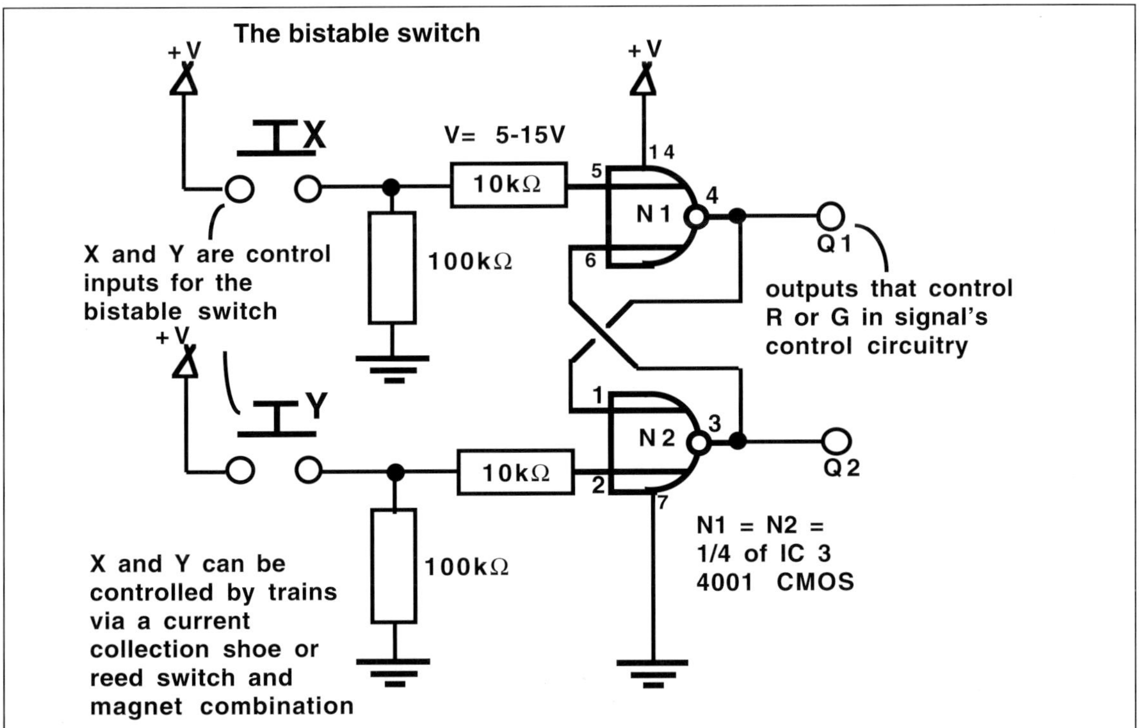

A short pulse at Y produces a high level state at Q1. Q1 remains at this high level even if the Y pulse is interrupted. Q1 goes low as soon as a short or long pulse is generated via X (when X is activated). Q2 always takes the opposite logic state of Q1. X and Y are not meant to be activated simultaneously. Full implementation requires an IC 4001 at each control module (a switch to each R and G input). Using blocking diodes, several units in succession can control one another so that the X and Y switches start and reset different control units along the rail line.

Modular sound generator for diesel locomotives

While real diesel locomotives are noisy beasts model diesels seldom make any noise at all. Here is a little modular sound system that can remedy this discrepancy. The more experienced among us will be able to pack the sound unit's components into an N scale engine. For most, it will be an accomplishment just fitting them into an HO unit.

If it were not for space restrictions, you could build this project on printed circuit board. In close quarters, you will have to solder component leads directly to each other. Note that the latter method can make trouble shooting difficult and is unsuitable for the novice. Patience is also an important resource for assembling the modules without a circuit board.

The system schematic shows a rate generator B connected to A modules. Use at least two A modules. Multiple A modules produce a better sound composite. The mixed A and B signals are led through a transistor to the amplifier input (see the sidebar on the TDA 2002 audio amplifier). For the sake of clarity, the volume trim potentiometer (10 kΩ) is included in the system schematic. It is a part of the amplifier module and need not be duplicated.

The rate generator is regulated by the locomotive's motor voltage with a circuit that includes a lamp and a LDR resistor (see the separate mini tip). The rate is matched to engine speed with a trim potentiometer connected in series with the LDR.

A problem with on-board sound is power supply. Track voltage is either very low or at zero when the locomotive is standing still. It then increases as the locomotive picks up speed. This brings about compromises in regard to sound. The main one being that the locomotive is silent when standing still.

A rather radical way of working around this is to install a 9 volt battery in the locomotive to power the sound generator (but not the LDR in the rate generator) and amplifier. It can be worthwhile to use a rechargeable battery if you operate frequently.

This book does not cover the advance digital technique (command control) that is now gaining popularity in the hobby. Just the same, I must mention how command control would simplify power supply for this sound unit. With command control, a constant voltage is always present at the rails, regardless if trains are running or standing still. A consequence of this is that model diesels emit sound both when running and when stopped. Only simulated motor rpm sound effect differs.

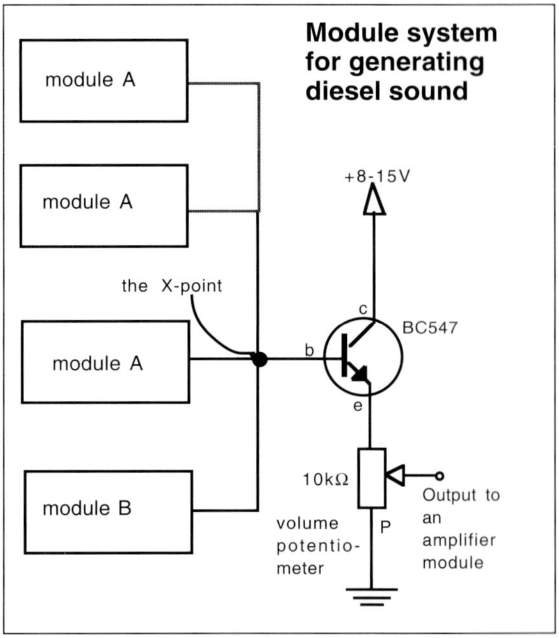

One of the classic american diesel engines here painted in the Utah colors. The model is common and well suited for experiments with lights and sound. The space between the power trucks can be machined to host a loudspeaker and inside the cover there is space enough for the electronic module.

Best result when in analog operation is achieved with onboard battery power. If using the sound generator in a digital system you can easily remote control the diselsound by letting a F-function toggle between "on and off" on the power supply.

Module for a sound generator in a diesel engine

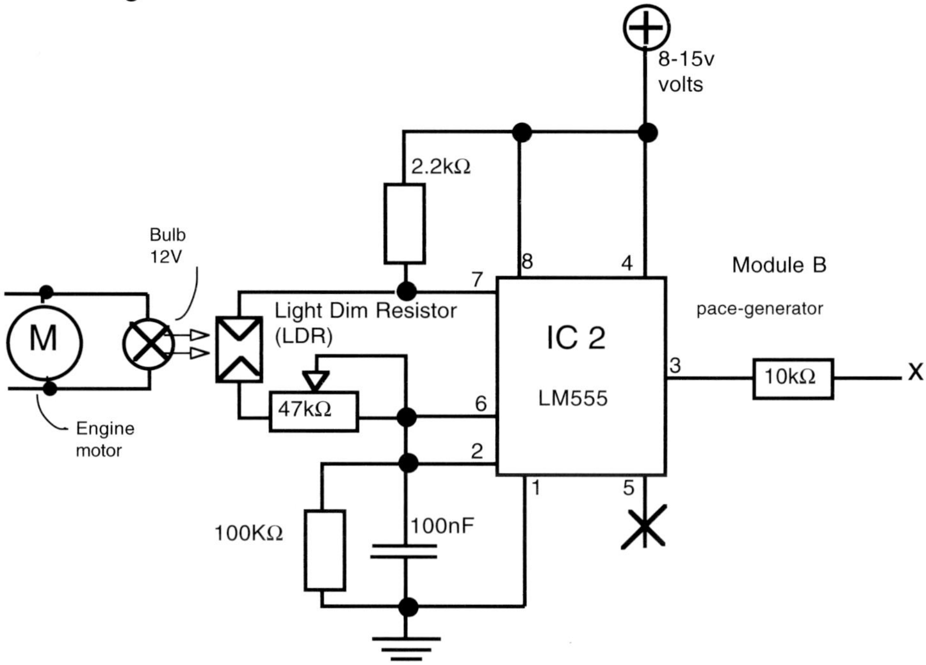

Module for sound generator in a diesel engine

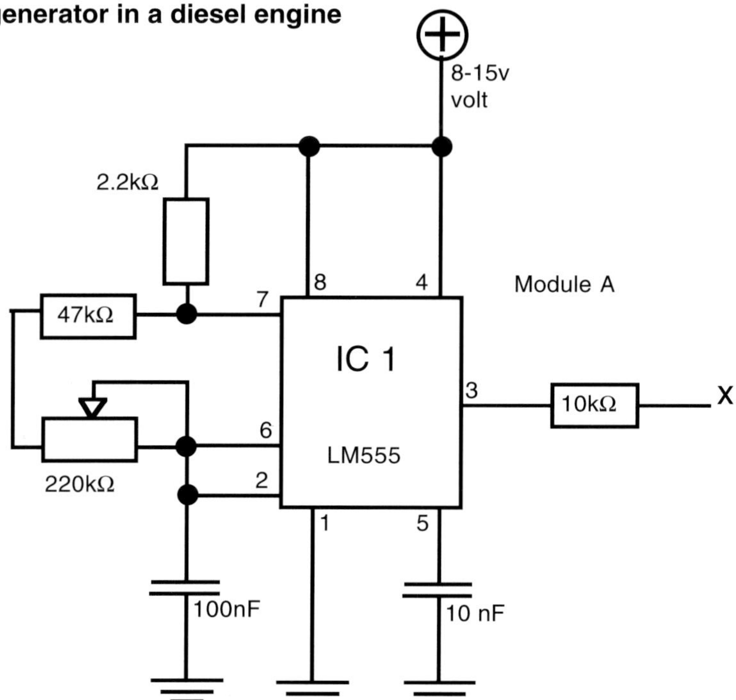

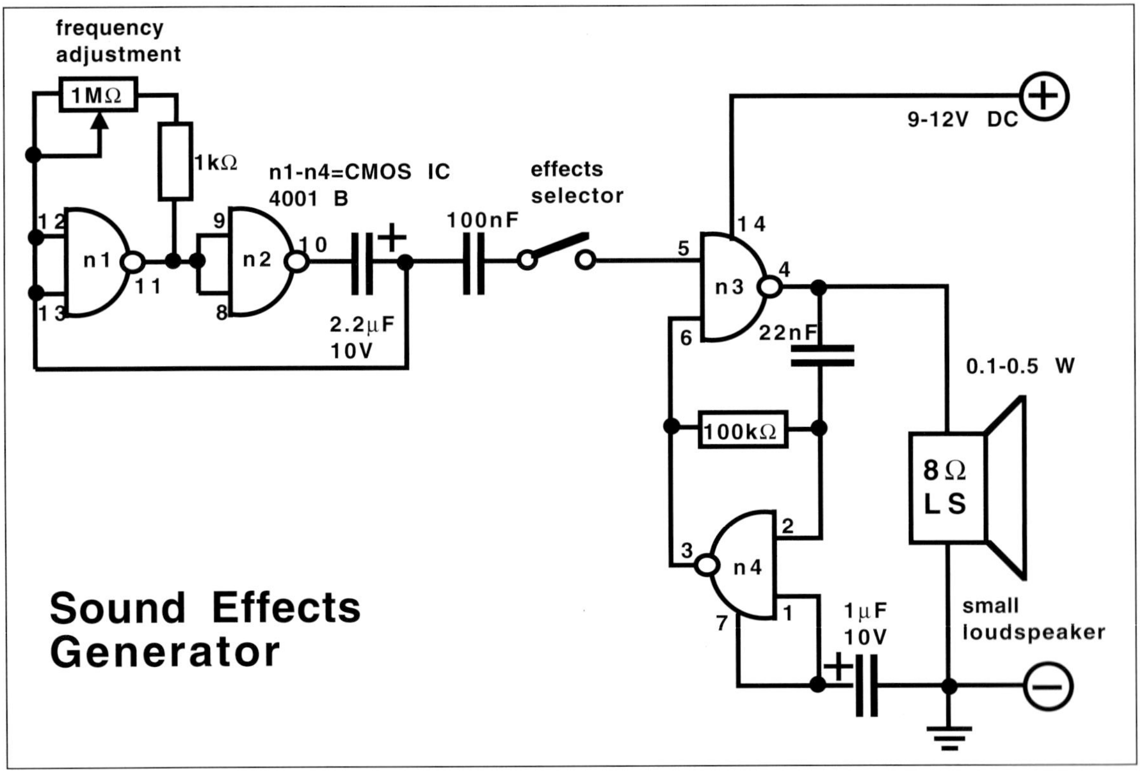

MINI-TIP

Sound Effects Generator

Don't expect this module to do sound miracles for you. However compared to its simplicity it does generate quite a few interesting sound effects. If you need the sound to be "bigger" the I recommend you to connect the audio amplifier module elsewhere in the book.

Starting at low frequencies we hit a lot of different engine sounds. Variants of horns, sirens and bird chitter are also present depending on the frequency adjustment.

By opening the selector you will start another sound facet, although strictly from an electrical point of view you are normally not supposed to leave a CMOS-gate (here pin 5) open. But this creates an interesting tropical sound that somewhere on a layout might be useful, like a garden restaurant.

This opens up a new adjustment possibility and you can change the 100 kilo ohm resistor to a 470 kilo ohms potentiometer in order to create more effects. With reference to the el-schematics, please note that the ground pin 7 of the IC is connected via a capacitor to ground.

PROGRAM CONTROLLED MODULES - WITHOUT A COMPUTER

We'll start with the basic module

In this construction project section, there is a slew of projects - 10 of them - that all share a common trait. They all make use of programmed control to create exciting and more realistic functions.

The title of this section gave you a good idea of what we'll be covering. We will be dealing with sophisticated program code, letting it control environments *and be controlled by outer factors*. Nevertheless, we will not use computers. This probably sounds puzzling - explanations are on the way.

The principle component in all of these projects is an extremely common IC. It is so common that I can just about say that it is a part of every computer. It's called EPROM (Erasable Programmable Read Only Memory) and is a memory circuit for permanent data storage. It doesn't forget its data even when the power supply is switched off. When used in a computer, the manufacturer usually loads it with general routines that handle everything from start-ups, cursor appearance and cursor blink rate, to how graphic images are displayed and letters interpreted. In short, everything that is the same at each start-up is good to store in an EPROM. In passing, I should also mention that there is a memory circuit called RAM (Random Access Memory) that loses its memory after each shut-down. This is where the user and the computer itself, stores all temporary data - data such as screen-displayed letter text, input data, etc. Theoretically, we could have used RAM for these projects but we would have to punch in control code each time we started up our layouts.

UNDER THE HOOD OF AN EPROM

A term that is sure to come up when talking about computers and their memory capabilities is "byte". A byte is a word length with space for eight binary "bits". This word length is recognized by the majority of semi-conductor manufacturers. A bit can have a value of zero (ground or "low level state") or one (+5 V or "high level state"). As long as we are on the subject, I should mention that computers generally operate on five volts. This is sometimes referred to as TTL voltage.

We store only signal levels that are at these extreme values of zero or one. For the computer to run fast, hops from one extremity to the other should be made as quickly as possible ("rise times" and "fall times" are terms that are used in this context). Home computers, as of this writing, run of speeds of up to 100-200 MHz (1 MHz is 1,000,000 clock cycles per second) and memory circuits must be capable of handling combinations of ones and zeros at these fantastic speeds.

When we come poking along with control technique working in the 0.05-10 Hz range, these memory circuits have it soft. On page 37 is a figure that shows how the eight bits in each storage address are arranged.

We will build electronic units that will slowly, and with plenty of opportunity for adjustment, step through these

During development, all program code was translated to assembler code in hexadecimal form and fed into the author's EPROM burner. The burner was actually an old Commadore 64 home computer with an accessory for loading code into EPROMs. If faulty code should inadvertently be entered into the EPROM, the entire chip can be erased by exposure to ultra violet light (watch you eyes) for 10-15 minutes. EPROMs erasers are available for purchase and are powered by household current.
Here you can see the various ICs used on the project. We'll be taking a closer look at them on the following pages.

SCEMATIC – MODULE A – Base unit

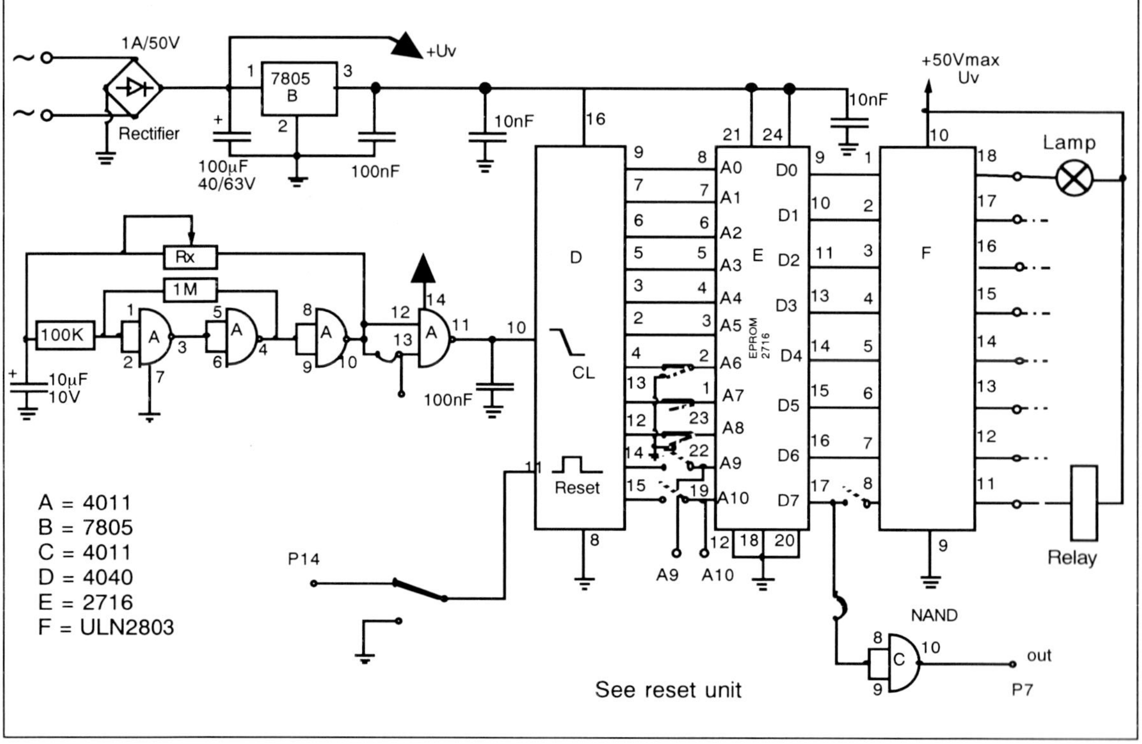

The base module A is utilized in all of the section's projects. That's why certain connections are unused in some projects and used in others. For information on the various connection points, see the diagram on page 66. The trim potentiometer Rx normally has a value of 1MΩ but at higher frequencies is changed to 100kΩ (2Hz-100kHz). Pin 13 on IC A is usually connected to pin 12. To stop the clock (that's to say, to directly control clock starting and stopping) for experimental reasons use this pin. One way to allow shorter programs to repeat is to ground the address pins of the higher values. Here, only A0-A5 are setup for operation.
The driver circuit F functions so that if D0, for example, is high (equal to 1), pin 18 on IC F will be grounded and the lamp (or in corresponding cases, a relay) will be activated.

addresses, one after the other. Program code means that we have in advance, loaded - called "burning-in data" - the ones and zeros of our choice in these storage places. During operation, all data outputs D0-D7 make number combinations available as output voltage levels. Here is a brief example of this. Let's say that D1 varies from one to zero at each step, just as shown in the figure. If we connect an LED to the D1 output, it will blink on and off (101010...).

For all projects, we will be using a type of EPROM that in the 27xx series. It has the advantage of being readily available and is offered by a variety of manufacturers.

We will use a chip size designated 2716 that can store 2 kilobytes (2 kB) or 16 kilobits (2000 x 8 bits = 16 kilobits).

BURNING AND ERASING

When you buy a new EPROM it is "empty". That happens to mean that there are masses of ones in storage. To put code in the EPROM, an EPROM burner is needed. There are several types available in many different configurations - both ready-to-run and as kits. Some are complete, self-contained units, while others are connected as an accessory to home computers. You can build one yourself for $20 or so. The world is full of EPROM burners and someone in your circle of friends may have access to one if you don't already have one yourself. If you exhaust all alternatives for obtaining an EPROM burner, drop me a line in care of the publisher and I will do what I can to try to help.

I would like to complement the picture by mentioning what should be done if code is burned (a "burn" is a short 25V pulse) incorrectly. There is almost no limit as to how many times an EPROM can be erased. This is something to keep in mind for those of you that buy surplus electronic components. Even loaded EPROMs can be erased. This is done with ultra-violet light (watch your eyes!) directed towards the EPROM's small, glass window. Erasing usually takes 10-15 minutes. Note that the entire chip is erased. You cannot just erase a portion of memory.

THE BASIC UNIT IN A CONTROL SYSTEM

In this first section, we will take advantage of the previously provided information and build a universal basic unit A. It will work with all other upcoming modules regardless of if we are programming structure lighting or operational mines. For those of you familiar with EPROM construction, you

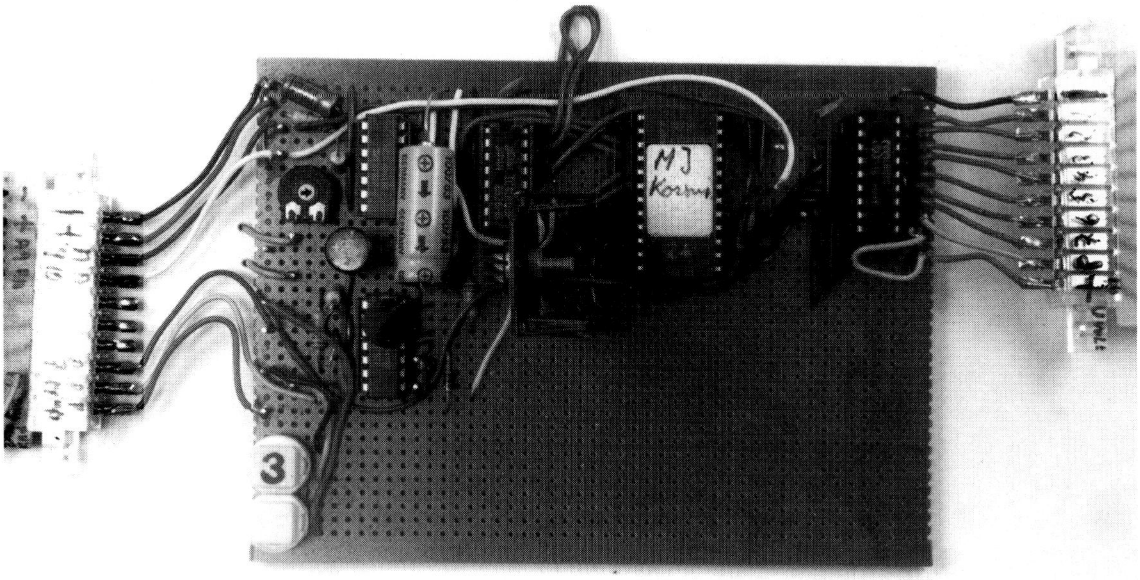

The photo is of the prototype module. The component layout figure below, showing the larger components and the trim potentiometer for oscillator frequency, matches the layout in the photo. It can serve as a guide in determining component placement on your own module.

can hop two steps forward and begin assembly with the component layout figure and electrical schematic as guides. The rest of us are going to take a closer look at EPROMs.

From the previous section, you should have a good understanding of moving through programs and how we would get an imaginary LED to blink. Many programming techniques are easy to apply if you just know that they exist. You can use these techniques when adjustments are made to a program or for writing your own programs.

We know that 2,048 steps are available (this corresponds to 2,000 when speaking of 2kB memory capacity). Let's say that a small program needs 30 steps. Do we really need to run all 2,048? No, we do not. Thirty-two is equal to 2^5. This means that to use 32 steps (this is as close to 30 as we can come), we need just five address lines in operation, A0-A4 (note, *five* address lines - we count from zero). The next binary steps are then programmed - all from A5 and upwards are grounded. This means that when the clock pulse searches for the remaining addresses in IC 4040 (D), addresses 0-32 are called again. In this way, we create an endless loop around a smaller portion of the total memory capacity.

If periodic reset pulses for other modules are needed, they are readily available from the EPROM. The figure showing the principle for program control provides an example of code used for a grade crossing. We could use data output D7 for reset control. We would program it as a "one" at all steps except the reset step. D7 would be a "zero" at reset and then fall back to "one" for the next operational cycle.

If more outputs than the eight outputs on a 2716 are

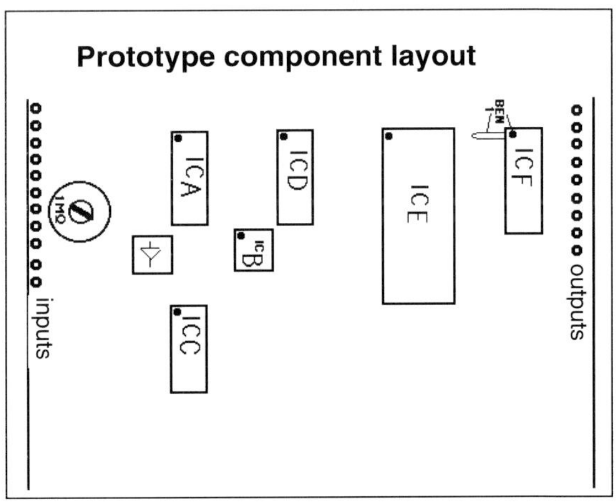

needed, two EPROMs can be connected parallel. This will provide us with 16 separate outputs. An effective use of double EPROMs is to connect them in matrices. Each matrice intersection provides a control output for a total of 64 outputs (8 x 8 = 64). It is the same principle as used on keyboards.

Besides the grade crossing program, we can have several other programs on the same chip and control their start only by coding the two highest address leads. There will be more on this in the next section.

As can be seen in the illustrations, the basic unit is rather uncomplicated. Hopefully, you have already built some of

Principles for program control

bell control (D3)

alternately flashing grade crossing lights (D1, D2)

EPROM address	$D_7 D_6 D_5 D_4$	$D_3 D_2 D_1 D_0$	
512	1 0 0 0	0 0 1 0	start address
513	1 0 0 0	0 1 0 0	
514	1 0 0 0	1 0 1 0	1 step per clock cycle
515	1 0 0 0	1 1 0 0	
516	1 0 0 0	0 0 1 0	
517	1 0 0 0	0 1 0 0	
⋮	⋮	⋮	
1023	1 0 0 0	0 1 0 0	stop address

of a total of 2,048 addresses, or storage spaces, with 8 bits in each (D0-D7) in EPROM 2716.
This example is from the grade crossing project.

Principles for program control

The 2716 memory chip can be compared to a large bookcase. It has space for 2 kilobytes, as indicated on the data sheet, and is equal to 2,048 bytes (or 8-bit words). Think of it as a bookcase with 2,048 shelves and 8-bit storage spaces on each. We'll now see what's on the shelves starting with shelf 512. This is where the grade crossing subprogram 2 starts in this example taken from page 60. The train is approaching the crossing, the red lamps are flashing in turn and the warning bell is ringing.
This is all controlled by various data bits at D0-D7. When the current bit is 0 (zero), the relevant lamp (or the bell) is turned off. When the bit is 1 (one), the lamp lights.
In this program, lamps D1 and D2 flash at a rate matching the rate used to step from address to address. Because the bell output at D3 only changes value every other address step, it rings at just half the warning light frequency. To slow the bell rate even more, let it keep its value through several address steps. An alternative to this is to lower EPROM frequency.

the easier projects in this book. I don't think that we need to dwell on construction technique. We are spared all problems associated with space limitations on this project. Even if your plans are for just a single program control module and you could compress the unit considerably, I would advise building the first unit with plenty of air around the components. It is always easier to trouble shoot when components are easily accessible.

MINI-TIP

THE BEGINNER'S VOLTMETER

To begin with, I would like to point out that this little instrument can never replace an analogue (with a pointer) or digital (with numbers) voltmeter or multimeter. If I don't have one of these handy or if I would like to have a DC voltage indicator built into a control panel, this circuit is an inexpensive alternative. The voltage range can vary with different component values.

In our circuit are five steps for approximated indication of 11V, 12V, 13V, 14.5V and 16V. While all LEDs could be the same color, using green, yellow, orange and red for rising voltages will ease reading. An alternative order would be to use green for the ideal voltage of say 13V, sandwiched with yellow and then red LEDs.
All resistors are rated at 1/4 watt and zenor diodes (with the double "tails") at 400mW.

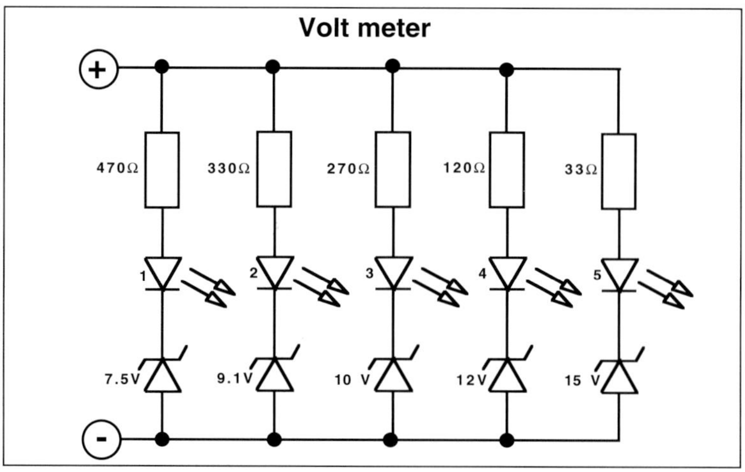

Lighting in houses and apartments

What is the motive behind this project? It's this - most structures on layouts lack lighting. And those few structures that do, are usually equipped with just one lamp, often lighting several floors of the structure. As a visitor, one is impressed with modellers that have managed to assemble the structure walls so as to avoid light leakage.

We are now going carry things a step further and make an ambitious attempt to mimic reality. And the methods are not especially complex.

Using mat board, we divide the structure interior into areas that will be conceived by the viewer as rooms. Keep in mind that structures with "finer" lighting will be objects for closer scrutiny. It can be worthwhile to add interior details. How much detailing is dependent on your patience with intricate, small format modeling.

DON'T BE BLINDED BY THE LIGHT

We'll return to electronics and concentrate our first efforts on house lighting. The accompanying lamp code (labeled "pgm 1") and floor plan are offered as inspiration. The lamps are connected to the basic unit, which is equipped with a driver circuit, IC F and can now be controlled via a reset unit as shown at the end of this project description.

You can operate your layout at *whole or half real-time.* This means that a day on the layout can have 24 hours or 12 hours. Most modellers prefer a considerably shorter day-cycle of 10-30 minutes. Others ignore these time-bound cycles completely but would still like to have lighting that reflects life in the real world.

The lamp code tries to imitate the different activities in a house during a dark, December day. When several structures are equipped with EPROMs, all running from the same program, lighting patterns will be repetitious. If you have the opportunity to program EPROMs, and have understood the explanation in the proceeding section relating to how code can switch lamps on and off, you can make an unlimited amount of "local adaptations". If you are unable to carry off such an advanced project, there is an alternative. Drastically vary the trim potentiometer adjustment so that the rise rate in the EPROM's memory address registers are unique from structure to structure. This provides a completely different feel of life and movement.

LET'S POWER UP THE APARTMENTS

Thanks to the computer, I was able to draw what I mean

LAMP CODE pgrm 1

D0—bath room
D1—bed room 1
D2—bed room 2
D3—kitchen/ceiling light
D4—living room
D5—stairway/garage
D6—hall
D7—exterior lighting

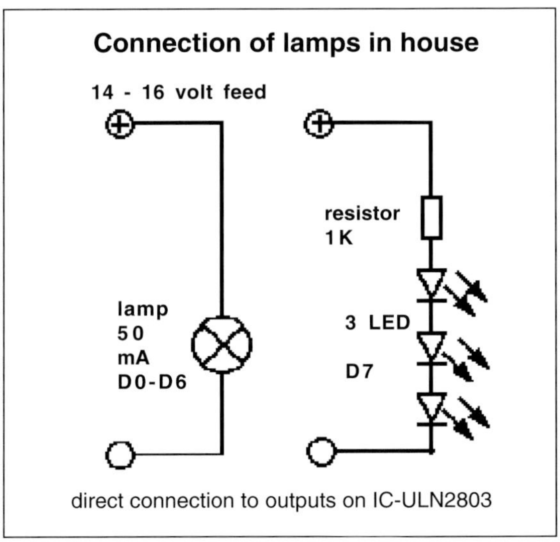

For lamps common to model railroad applications, appropriate feed voltage is in the 14-16V range. For LEDs, use 1kΩ series resistors.

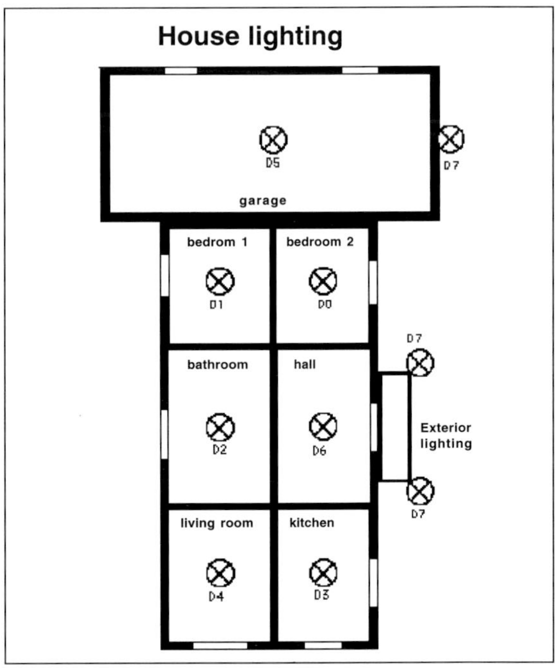

This is my prototype plan showing a two bedroom home with a connecting garage. To use the provided program, your model home lighting setup should be similar.

Lighting in an apartement house

This figure illustrates the imagined apartment house where the day's 24 hour lighting cycle is compromised to 10 minutes or perhaps an hour or so. Cycle duration is adjusted with a trim potentiometer on module A. If the provided code is to be used, lamp connections should be similar to those shown. The program reflects various nocturnal habits of the inhabitants and includes code for lighted signs and street lighting.

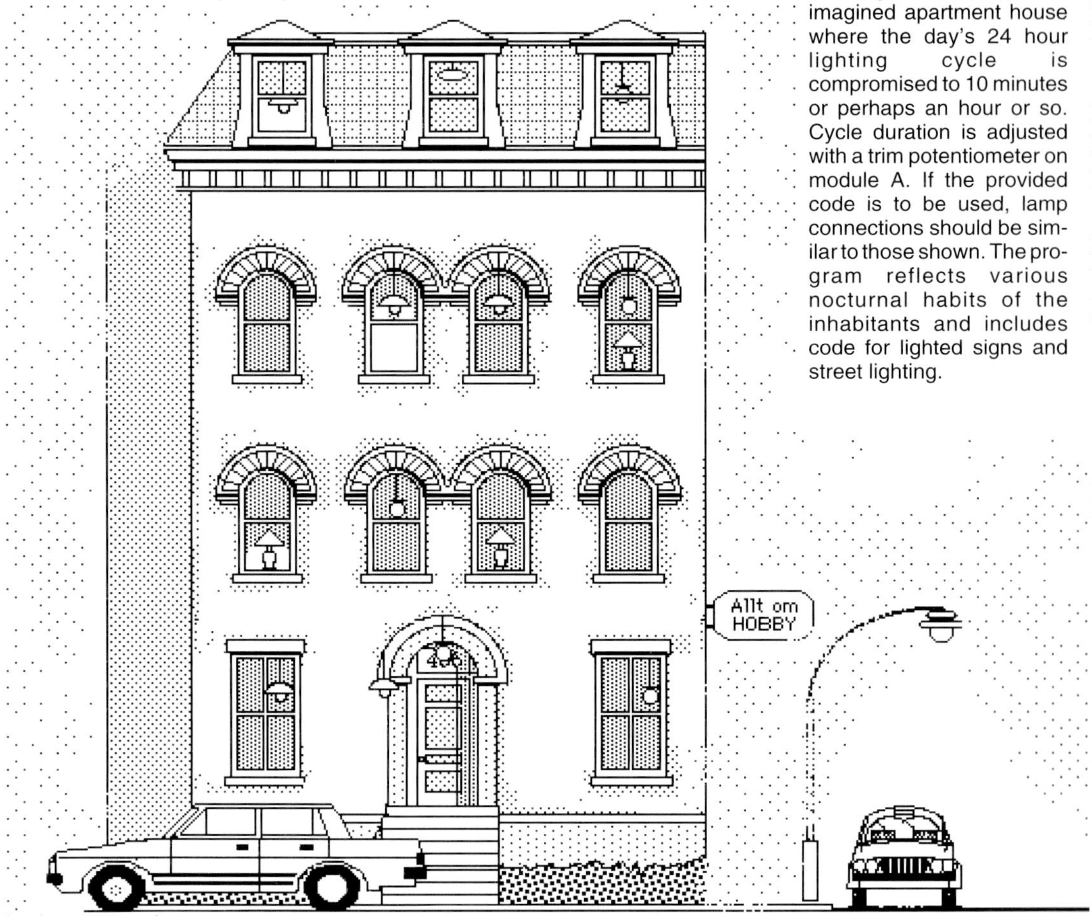

EPROM house – program 1, One floor house

Source code in running order

80	80	80	80	80	80	C0	E0
E0	F0	F0	E8	E8	EA	EB	EA
E8	E8	C8	C8	C8	C8	D8	D8
D8	D8	D8	D0	D0	D0	D0	90
90	90	90	90	90	90	98	98
98	98	98	98	98	D8	FA	FA
FA	FA	F1	60	44	44	44	44
06	07	07	07	07	06	02	02
02	02	02	02	02	02	02	02
02	02	02	02	02	02	02	02
00	00	00	00	00	00	00	00
00	00	00	00	00	00	00	00
00	00	00	00	00	00	00	00
00	00	00	00	00	00	00	00
00	00	00	00	00	00	00	00
00	00	00	00	00	00	00	00

program 2 – multifloor, city apartement house

00	80	A0	A0	A1	A1	E9	E9
E9	C9	C8	C8	CC	CC	CE	CE
CA	CA	CA	CA	CA	CA	CE	CE
CE	EE	EE	EE	EE	EE	EC	EC
EC	EC	EC	EC	E4	E4	E4	C4
D4	D4	D4	D6	D6	D6	96	97
97	97	97	97	95	95	95	95
95	95	95	95	91	91	91	11
11	11	10	10	10	10	10	10
10	00	00	00	00	00	00	00
00	00	00	00	00	00	00	00
00	00	00	00	00	00	00	00
00	00	00	00	00	00	00	00

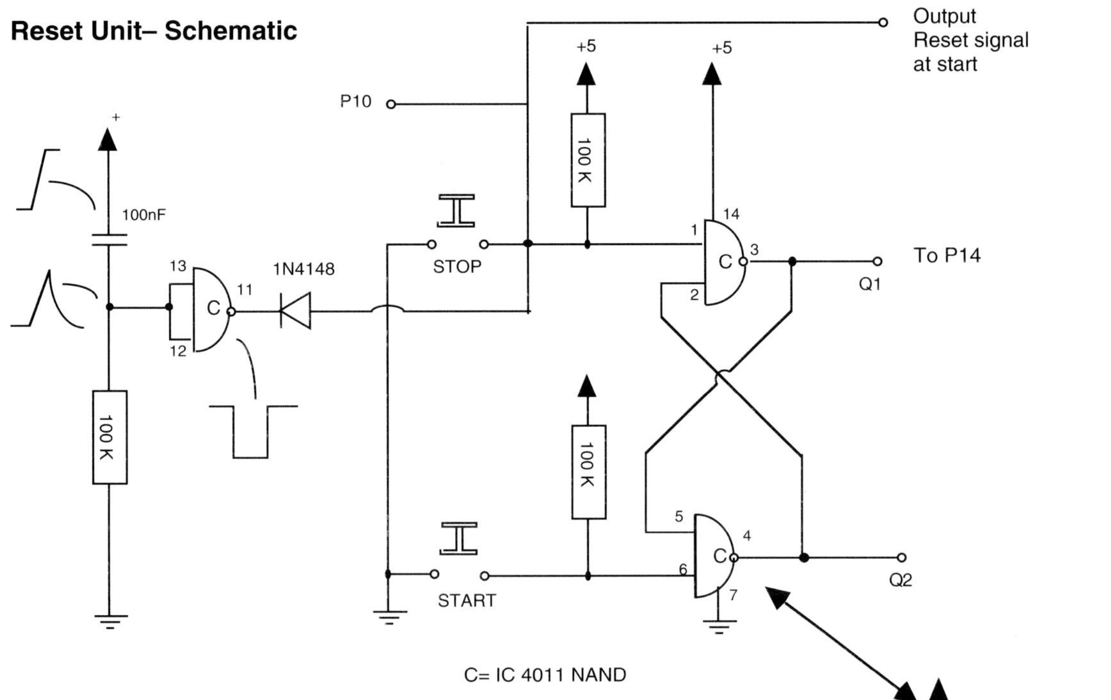

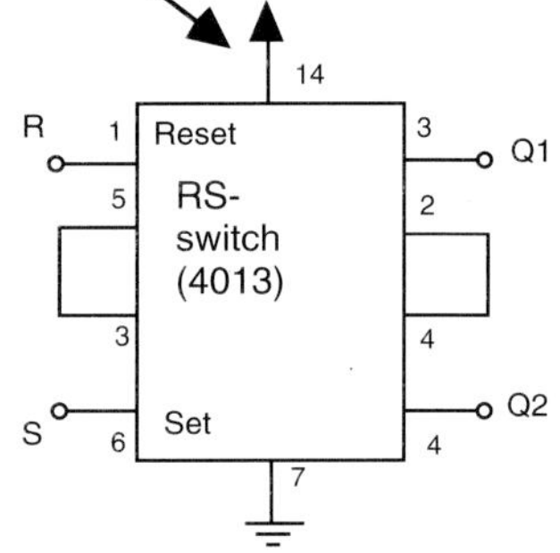

For structure illumination, it can be good to have the option of manually starting and stopping program control. This is because changes to output combinations or clock frequency may require adjustment to the program advance rate. The reset unit can be built using either an IC 4013 or IC 4011, depending on what is on hand.

LAMP CODE pgrm 2

D0—apartment on 3rd floor
D1— apartment on 3rd floor
D2— apartment on 3rd floor
D3— apartment on 2nd floor
D4— apartment on 2nd floor
D5— apartment on 1st floor
D6— signs and show window
D7— street and entry lights

with an active layout, apartment house. There are lighted signs, shop window displays and street lighting too. There are also a number of apartments populated by people with varying nighttime habits and lighting cycles must be varied. My biggest problem with this project was that the total number of lighting outputs was insufficient for a structure of this size. In the proceeding section, simple ways around this problem are described.

I'd like to take the opportunity to squeeze in a few more tips that you may find enticing. An electric smoke unit in a chimney that periodically emits smoke, rising towards the sky is one. This is more than what IC F can handle. Between IC F and the smoke unit, a miniature relay should be added.

In an earlier project book, I presented a sign lighting module with running lights (LEDs). One of these modules could be wired in as an attention getting feature during the "evening hours".

As with the house lighting project, covered earlier, I have also put together a suggested lamp code for apartments. It is labeled "pgm 2".

As can be seen in the earlier figure with all of the memory circuits, I have chosen to collect these programs in EPROM-HOUSE. How an EPROM can be divided to store several sub-programs is described in the grade crossing project description towards the end of the book. Program numbering follows in all cases that prototype (a maximum of four programs with 512 steps).

MINI-TIP

Light control in smaller buildings

The full fledge light control for any timing and complex lighting is found in the chapter of eprom control in this book. This is a more limited but easy to build module.

We use the classic timer IC 555 and this time it operates extremely slow from time cycles of ca 1 minute to half an hour.

Looking at a smaller house from outside the interior lighting is mostly on during afternoon and the evening. So we put room lighting in 3 different locations and let them be controlled via a walking ring IC (4017). There are totally 10 output possibilities from this IC but here we only use 3 (your choice could be to use more output control pins).

The intelligent building

(changing room lighting)

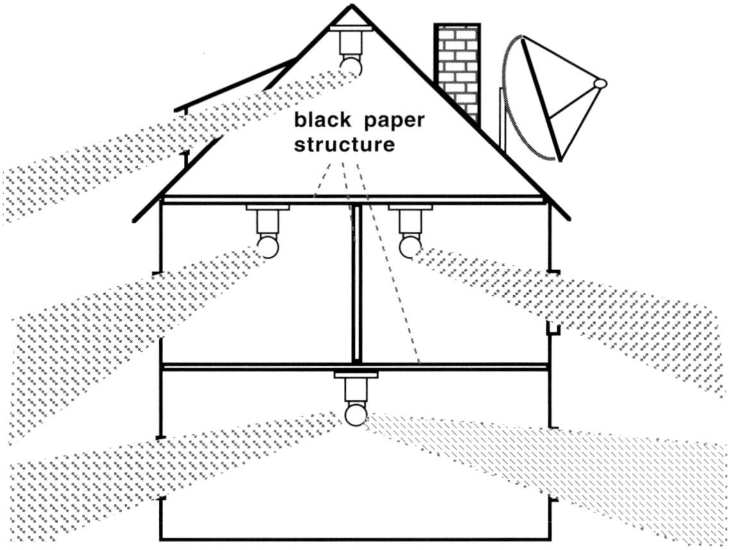

So the left bulb is in the kitchen and will be on during 2 out of 10 cycles. The middle bulb is in the garage or bathroom and will only light up once (2:nd cycle). And finally the right bulb in the living room first lights up during 1:st cycle and then during cycles 3 and 4. Here is the correct pin sequence if you want to change order of lighting or increase number of bulbs: 3-2-4-7-10-1-5-6-9-11. If you use the last pin the connect pin 15 to ground. The length of each cycle is adjusted via the potentiometer .

Elschematics for smaller lightcontrolled building

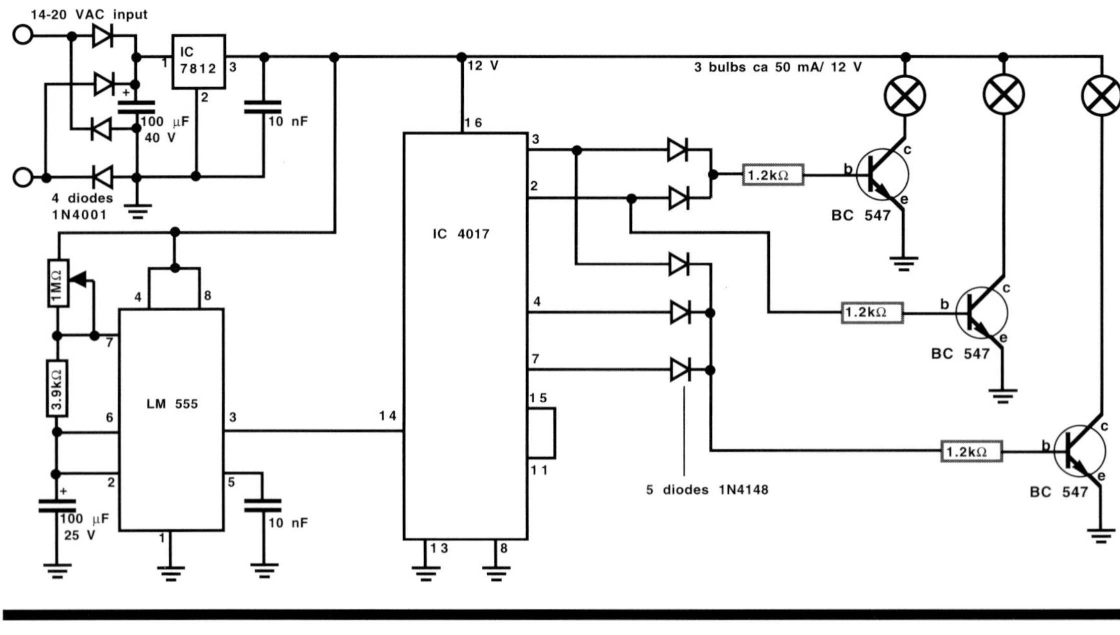

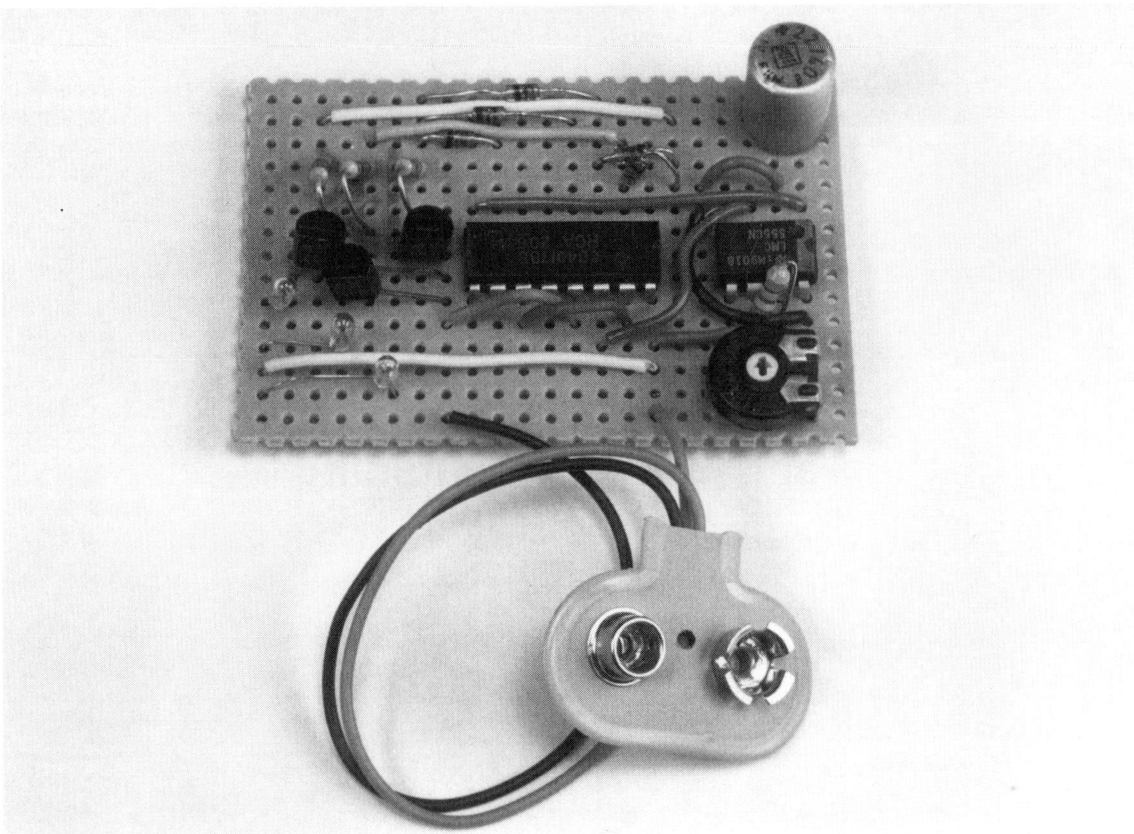

MINI-TIP

A horn for diesel locomotives

This is a small, uncomplicated unit built around a single IC, 4011. Three of the four gates are used. If you don't intend to use the fourth gate, connect its inputs to either plus or ground. There is a separate mini-tip sidebar that covers CMOS gate connection.

The 10kΩ trim potentiometer is for adjusting sound. The circuit is intended to work with a small TDA2002 amplifier. This amplifier also has its own mini-tip sidebar.

Choice of gates in the 4011 is up to you. Take a look at the schematic. The gate at the far left sounds the horn when both of its inputs are connected together. Because the circuit was designed for on-board use, we need a way of controlling it remotely. The schematic shows how a reed switch mounted under the locomotive activates the horn when it passes either a permanent magnet or electromagnet. Three obvious locations for magnets are at tunnel entrances, just before grade crossings and on station approach tracks.

Diesel horn

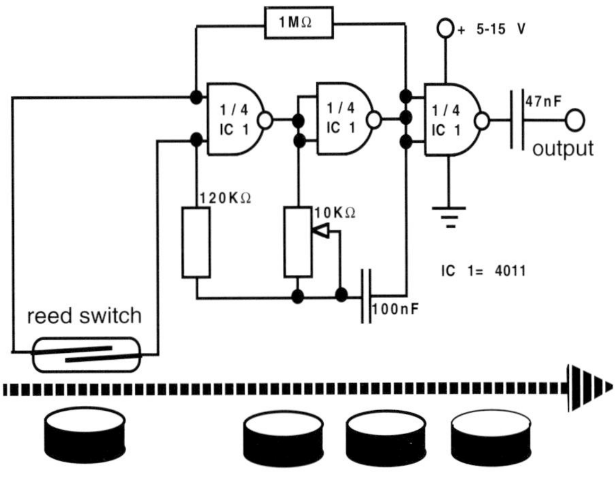

The time controlled street intersection

If you are looking for that special project, this may be the one. Investment in relation to pay-back is moderate. In the first book was a relatively simple, but at the same time quite serviceable, project featuring motor vehicle traffic control. It regulated red and green in constant, opposing positions for the intersecting streets and gave a yellow indication in all four directions simultaneously, loosely following Swedish practice.

At the time, "we" couldn't do better. With programmed control at our disposal, we can now simulate a specific intersection with correct detailing - from pedestrian signals to nighttime, flashing yellow signals. I stopped at a signal-regulated intersection close to my home and jotted down the signaling sequence. This resulted in the Diagram for Light Switching. As you can see, a complete cycle takes 31 seconds. This intersection is pictured in the figure, Intersection1. Keep in mind that the Swedish signaling system differs from the American. Yellow is used in combination with red and green for transition indication.

A signaled intersection can be an important center of interest on any layout. Model figures should be placed in groups - the effectiveness of the scene is often in proportion to the total number of "small folk" invested in. A legitimate criticism of a model intersection is that it lacks sound.

Diagram for light switching

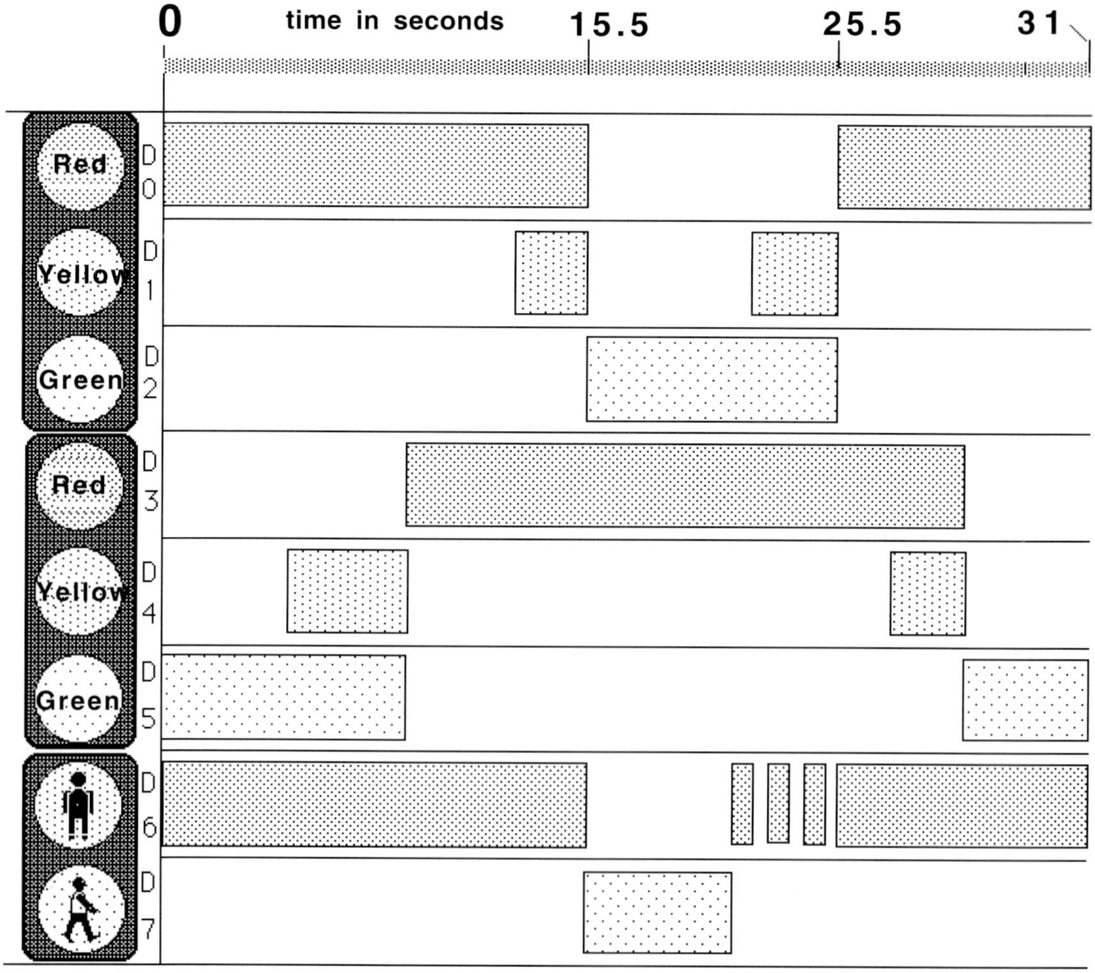

The diagram reflects operations on the first intersection project.. Using the diagram as a guide, regulation of each lamp was determined and corresponding program code written.

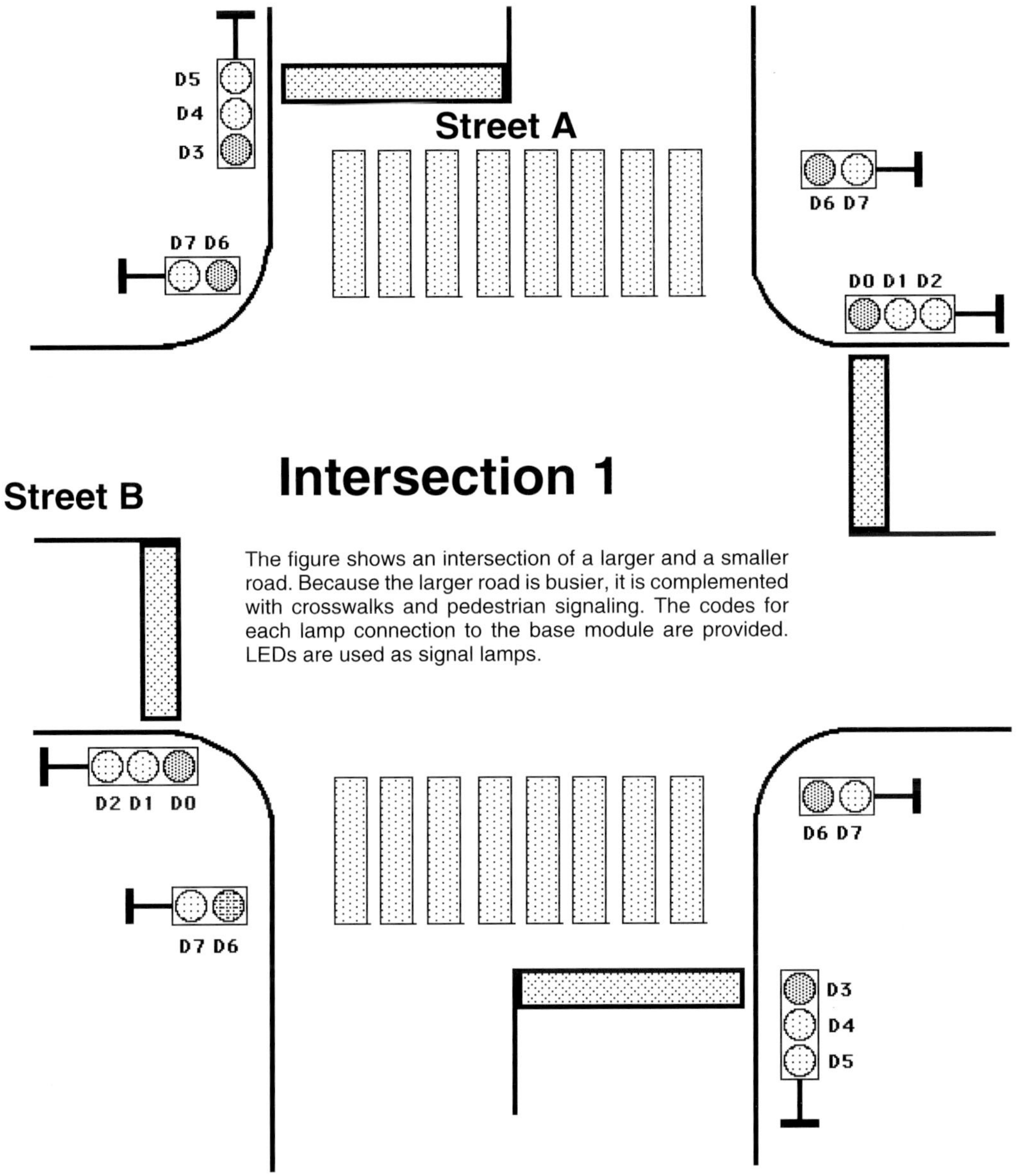

Intersection 1

The figure shows an intersection of a larger and a smaller road. Because the larger road is busier, it is complemented with crosswalks and pedestrian signaling. The codes for each lamp connection to the base module are provided. LEDs are used as signal lamps.

There should be a tumult of sound - a mix of blaring horns, the sound of tires on asphalt, accelerating motors and the like. A small cassette player, playing authentic roadway sounds can be a good remedy for silent, model intersections.

AN INTERSECTION WITH VARIATIONS

The first example was of a rather quiet, suburban street intersected by another, even quieter. It had its own control program called Intersection 1. Let's shift are attention to the big city, where major thoroughfares cross one another. A more equal distribution of green indication is needed between the crossing streets and all corners equipped with pedestrian signaling. The figure labeled as Intersection 2 illustrates this. This is a typical example of compromise in that what we really need is at least two more control

EPROM INTERSECTION, program 1

Intersection on main road with signaled regulation of cross street. Source code is in order of operation.

61	61	61	61	61	61	61	61
61	61	61	61	61	61	61	61
61	61	61	61	71	71	71	71
49	49	49	49	48	48	48	8C
8C	8C	8C	8C	8C	8C	8C	8C
8C	8C	8C	8C	8C	4C	0C	4E
0E	4E	0E	49	49	49	49	59
59	59	61	61	61	61	61	61
61	61	61	61	61	61	61	61

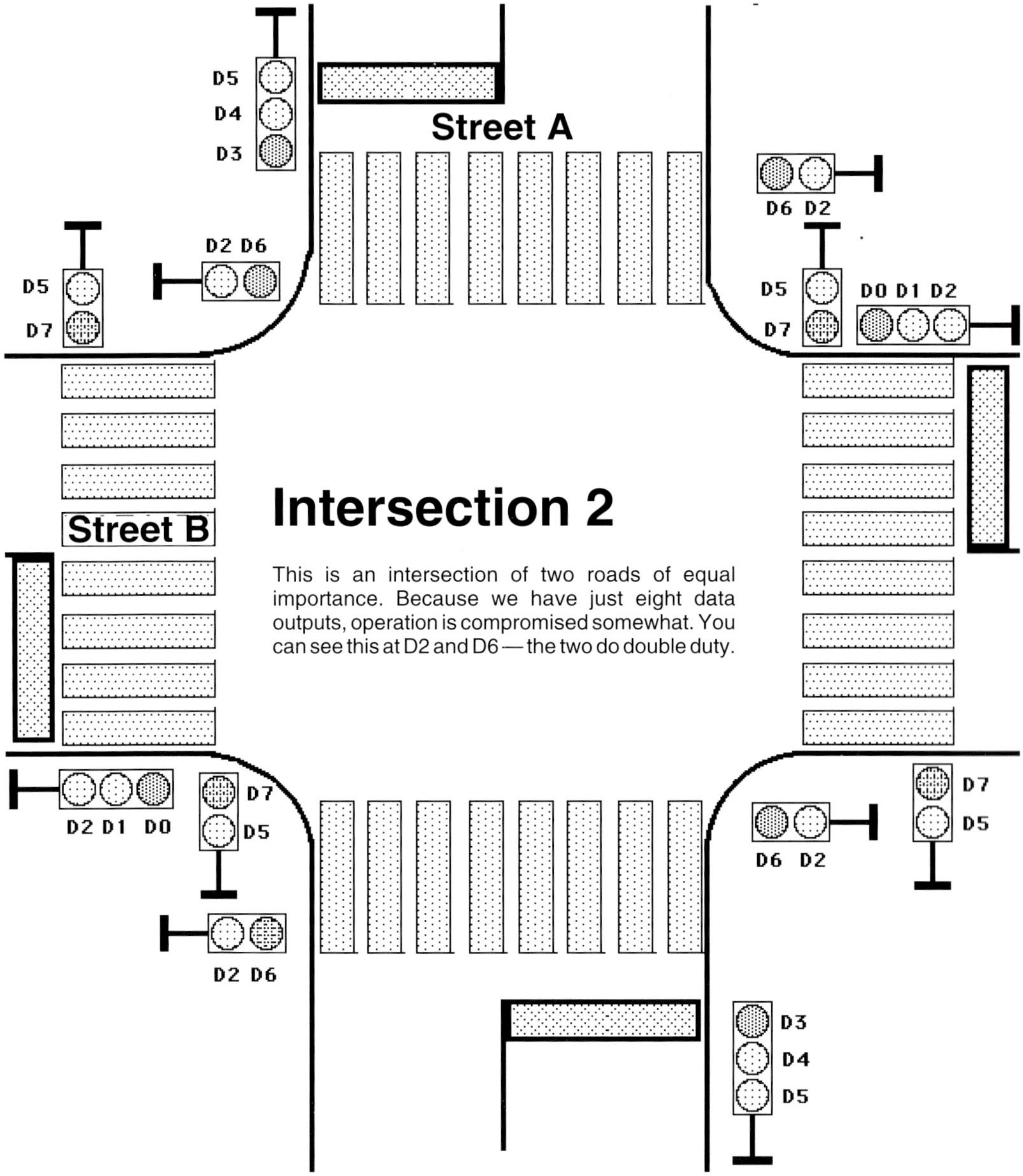

Intersection 2

This is an intersection of two roads of equal importance. Because we have just eight data outputs, operation is compromised somewhat. You can see this at D2 and D6 — the two do double duty.

outputs. For now, we'll limit ourselves and above all, the program code, to the eight outputs on an EPROM. That gives us an intersection that matches the figure.

As long as LEDs of the same size are used, all three of the example intersections are wired in the same manner. A tip for those of you modeling in N and Z scales: the German company, Busch has nice, small-format signals. They are well made, and inexpensive when quality is considered. Of course, if you want even more for your money, you can construct the signals yourself. In consideration of scale, 3-5mm is suitable for HO and O scale while 2mm for N scale. The smallest LEDs are 1mm and are suitable for both N and Z scale. If you will be using the smaller LEDs, you should be well experienced with soldering technique...a little swearing now and then may also be required.

EPROM INTERSECTION, program 2

Intersection featuring four-way pedestrian signaling. Source code is in order of operation.

61	61	61	61	61	61	61	61
61	61	61	61	61	61	61	61
61	61	61	E1	61	F1	71	F1
71	C9	C9	C9	C9	CB	CB	CB
8C	8C	8C	8C	8C	8C	8C	8C
8C	8C	8C	8C	8C	8C	8C	8C
8C	8C	8C	8C	CE	8E	CE	8E
C9	C9	C9	C9	D8	D8	D8	61
61	61	61	61	61	61	61	61

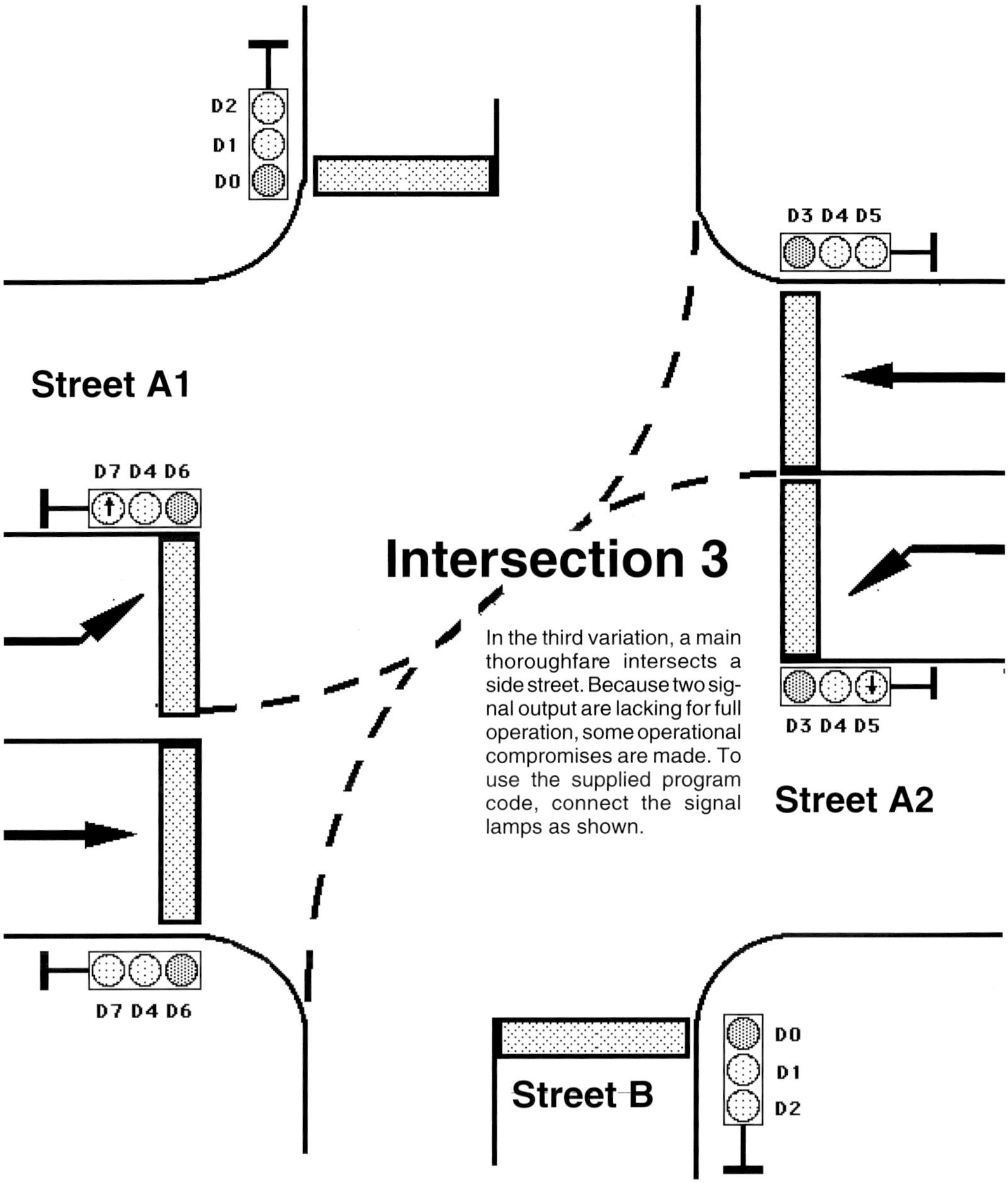

In the third variation, a main thoroughfare intersects a side street. Because two signal output are lacking for full operation, some operational compromises are made. To use the supplied program code, connect the signal lamps as shown.

Although I've never seen a layout featuring a major highway, this hasn't stopped me from providing the program shown in the figure, Intersection 3. While planning the previous two programs featuring suburban and city intersections, it struck me these should be balanced by a major highway intersection, complete with left turn lanes. As you may have guessed from all of the signal lamps involved, minor compromises were made. But these are marginal and the overall effect on the viewer is positive.

The German model accessory company Noch, offers nicely done roadway surface in HO, N and Z scales. It is sold in one meter rolls, is elastic (meaning that it can be used to model curved roadway) and self-adhesive - you only get one try. And lastly, the traffic signs that "belong" to intersections can be clipped out from road maps.

EPROM INTERSECTION, program 3

Intersection on main road with signaled, left turn lane. Source code is in order of operation.

4C	4C	4C	4C	4C	4C	4C	4C
4C	4C	4C	4C	4C	4C	4C	4C
4C	4E	4E	4E	49	49	49	61
61	61	61	61	61	61	61	61
61	61	61	61	61	61	61	70
71	71	71	89	89	89	89	89
89	89	89	89	89	89	89	89
89	89	89	4B	4B	4B	4C	4C
4C	4C	4C	4C	4C	4C	4C	4C

MINI-TIP
Road traffic control

This is the electronic platform for any motion and running light you might need on your layout. It might be a road traffic warning as suggested here; or advertisement around a shop window, etc. Use Lettraset to make the sign or if you have a desktop facility, simply design the stand according to your local infrastructure or traffic management via software and plot or print it.

Basically we use one IC (gates n1-n4) to make an oscillator or pulse generator and the second to distribute the pulses in a walking ring order. A simple way to power it is using a 9 volts battery. If you want to use the existing 14-16 V AC power available at your layout, then omit the battery. Instead use a bridge rectifier to get the DC and a Capacitor (220 µF / 40 V) to filter the DC. If depending on the feeding from your transformer you then measure a higher voltage than 15 V, then use a voltage converter such as 7812 to limit supply to 12 volts.

Road traffic control

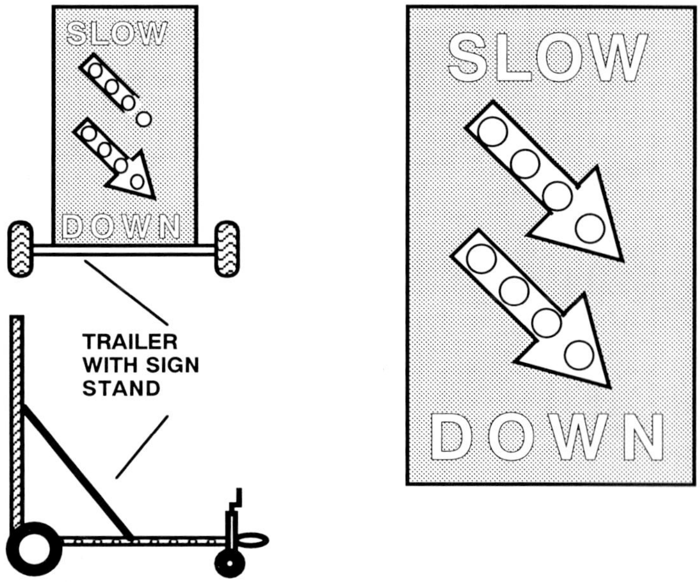

Road traffic control — elschematics

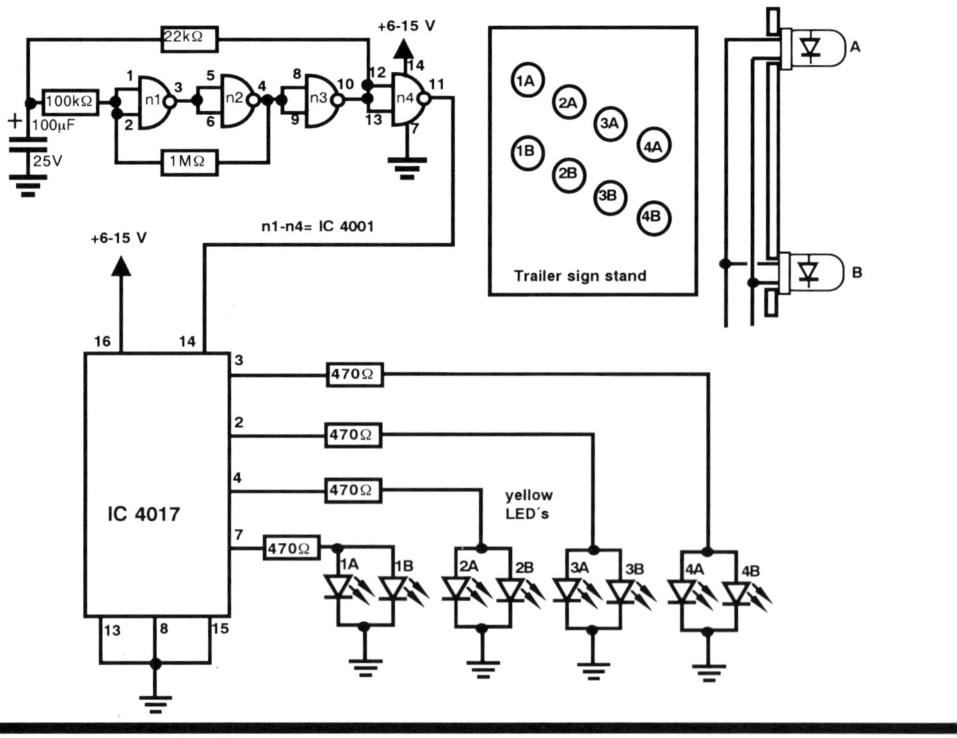

MINI-TIP

General flash warning

This easy to build module can be a useful "platform for many different warning purposes. Around a Heliport on a building roof is one obvious example. A high broadcasting transmitter antenna that is a danger to the air traffic is another.

The pulse generator is based on the timer IC 555 and has here a ready set frequency, a little faster than 2 flashes per second. Through increasing the value of the top resistor (10k ohms) you will lower the flashing pace when using it for other purposes.

There are two alternating outputs generating a double sequence of flashes. In order not to overload the pin 3 output pin we here use a PNP transistor to power the second part of the LED´s.

General Flash Warning lights

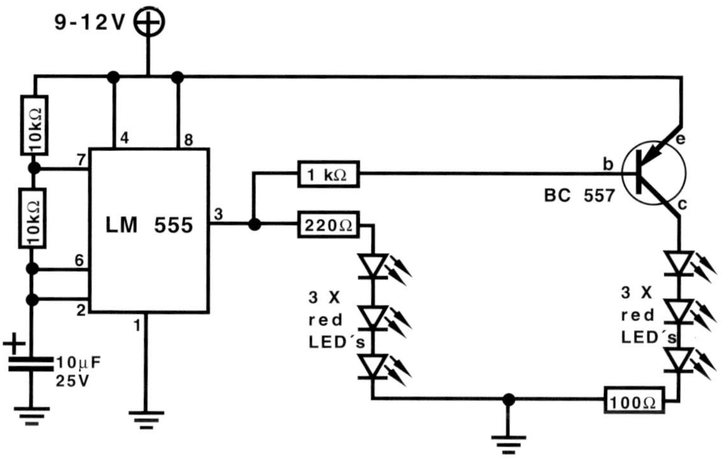

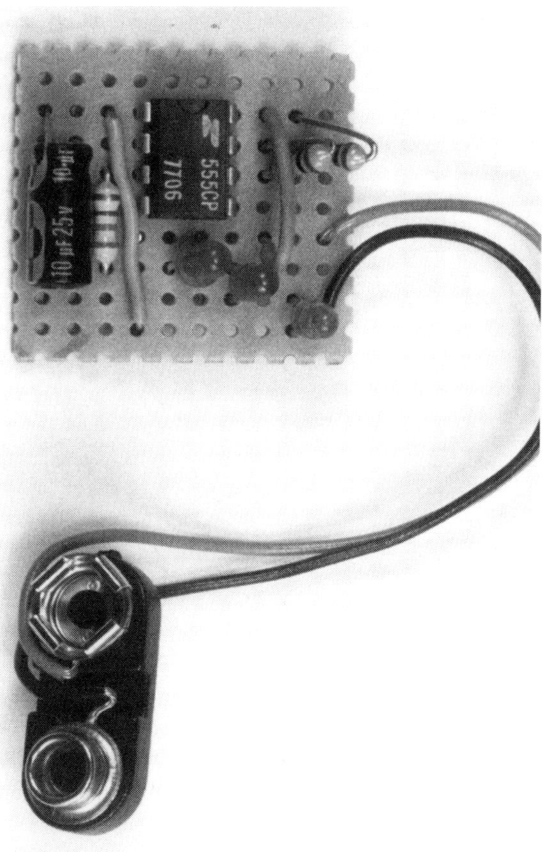

Protype for general warning.

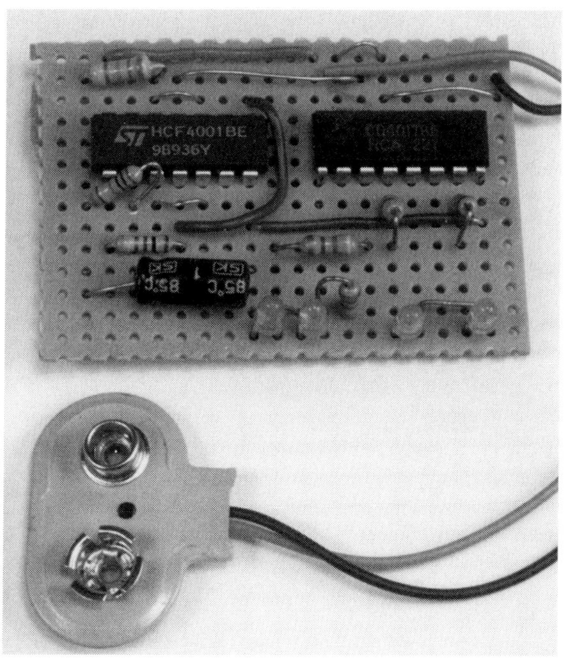

Prototype for road trafic control.

Three program controlled elevator projects

Up until now, programmed control has only concerned regulation of lighting and in one case, sound. We'll now raise our sights some and add motion control to our repertoire.

The introduction of the elevator project makes mechanical construction skills a must. We're entering an area where major model manufacturers have yet to venture. This makes the project both more exciting and guaranteed free from annoying plagiarism on your neighbor's layout.

I'll open by describing a bit of the project's mechanics, drive train and safety features. I've built quite a few elevators over the years similar to the illustrated projects. Naturally, I have tested several different drive train details and used varying connection methods. What follows shouldn't be taken as assembly instructions but rather as tips based upon my own experiences. In practice, the most decisive factor in construction design is the content of your "treasure chest" - that's the one the wife calls a junk box and that the vacuum cleaner just *happens* to empty now and then. If you come across a small, compact electric motor with low gearing and preferably, a long extending axle, many of your problems are already solved. It's worth its weight in gold.

Diverse small spools of thread, string, weights for placement in the elevator and the like are very useful in this type of construction. The elevator shaft, or open framework on a construction site elevator, can be modelled in many different ways. I like to use balsa sheet with bracing for shafts and "hard-soldered", thin square brass stock for open frameworks. Driving is easiest using a continuous line as shown in the figure. It provides steady action and the risk that the elevator tips and fastens in the middle of the shaft is minimal.

The emergency stop function consists of two independent systems. The guiding principle is that the elevator should never reach its end positions. The control program should stop the elevator, not the emergency stops. Don't forget that there is an additional adjustment via a trim potentiometer on the basic module that has a direct effect on elevator travel distance. The first emergency stop system is shown in a separate figure and employs reed switches. It is usually enough with this circuit. The second system provides double insurance from mangled elevators. Run the motor power feeds to the elevator shaft. Connect them across the shaft with either a thin length of silver wire or leaf spring contacts. Mount the wire or contact beyond the ordinary stop position. If the elevator should pass its stop position, power feed interruption is handled mechanicaly.

A CONSTRUCTION SITE ELEVATOR

The modeled construction project is especially rewarding because it provides the layout with life and movement. By construction project, I mean roadwork, a house being roofed, etc. For larger buildings under construction, elevator shafts usually feature an open framework and are placed on the outside of buildings. We can model the framework in brass. The first figure illustrates a construction site elevator. At the ground floor is a short, safety gate of the same type used at minor, railroad grade crossings. It and the lamps at each floor are controlled by the program. The elevator can be made of mat board, painted and equipped with a small weight and magnet. For more detail, use brass stock construction - in brass, you can even model the safety railing.

A MINE CROSS-SECTION

A mine as seen from the surface is often non-descript (we'll ignore open-pit mines for now). Some mines don't even have the towering elevator shafts that would otherwise indicate their underground activity. An example that comes to mind is a large copper mine in northern Sweden. The only visible surface signs are the shaft opening and a small framework structure.

Yes, underground mining operations can be difficult to model on a layout. This is how I have tackled the problem. The layout is supported by a framework with sturdy legs at the corners. The mine is at one of the layout ends. The complex's surface level is approximately four feet from the floor. From the four foot level down to about one and a half feet above the floor, operations can be viewed through a vertical cross-section. Some drifts are exposed along their longitudinal axis while others are "clipped" straight across and you can see deep into the passageways.

A few words on tunnel construction are in order. I used shoe boxes and tin foil, crumpled, straightened and painted black. Dark, matted colors make for good rock imitations. From the surface level, hidden trackage is spiraled down to the lower levels. It is quite unsettling to see the little mining train from the surface suddenly appear in the lower drifts. Water seeps along the rock faces and collects in a basin at the bottom of the mine. A pump brings the water back up.

In this dimly lighted environment, where activity is just barely visible, an operating elevator is the finishing touch. An elevator in a sterile setting doesn't add much to the layout. You have to put a little work into the surroundings too.

VERTICAL ORE TRANSPORT - AN UPLIFTING STORY

We have railroads to run and must delegate as much mining activity as possible to programmed control. So we'll stay in the lower regions a while longer and take a closer look at a special, automated transporter for ore. The accompanying figure will show you what I mean. The lifting bucket resembles a dump-car and is filled with ore using a remote controlled, electromagnetic valve. It is then raised to the upper position where a protruding stop causes the bucket to tip its load into the ore bunker. After discharging its load,

Construction site elevator

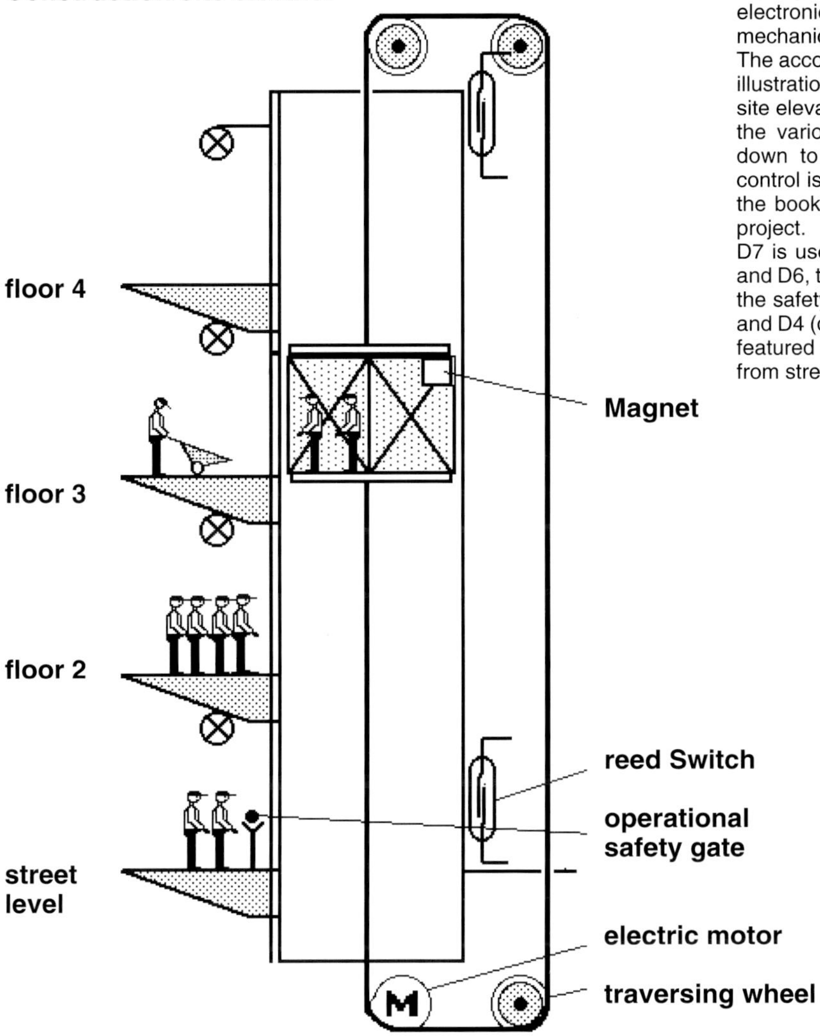

In that project description content is primarily electronics- orientated, I won't dwell on mechanics and other such important topics. The accompanying attempt at architectural illustration is meant to depict a construction site elevator. The elevator travels between the various floors and often makes trips down to street level. Electrically, motor control is identical to the next to the last of the book's projects — the grade crossing project.

D7 is used to move the elevator upwards and D6, to move it downwards. To regulate the safety gate at street level, use D5 (up) and D4 (down). Flashing warning lights are featured at each floor. These are regulated from street level.

it returns to the bottom position and repeats the sequence as many times as desired. I used a cheap Märklin dump-car from a beginner trainset as a lifting bucket. You can usually find similar cars on your dealer's used equipment shelf just because they're considered too "simple". They are ideal for our purposes. The ore bunker is an upside down, 16 ounce plastic bottle with the bottom removed. As stated in the figure caption, the electromagnetic valve is from Arnold.

Scene framing for this elevator project is very important. Be careful to hide certain parts of the elevator construction from view. Use plenty of tin foil, sprayed in a dark color to imitate rock face. If you want to achieve something a little extra, a miniature speaker, tucked away in a drift can do wonders. You could, for example, tape echoing voices, water dripping, etc. There are plenty of similar sounds in actual environments that can be taped.

For the ore elevator, I should mention how the connections to the basic module's outputs are made in relation to program code. If you have a warning light or buzzer at the lower position where the bucket is filled, connect it to D0. The corresponding light or buzzer in the upper position is connected to D1. The electromagnetic valve in the figure is activated by D2. Winch house or mine office lighting can be controlled by D3 and pit entry lighting by D4. There is an additional output for an electromagnet device at the upper termination position (D5) for alternative unloading control of the ore bucket.. Upward movement of the elevator is handled by D6 and downward movement by D7.

Connect LEDs to the outputs for initial testing and dry runs to check that time sequences are correct. Don't forget that both times and elevator travel distance can be adjusted with the trim potentiometer on the basic module.

Mine elevator

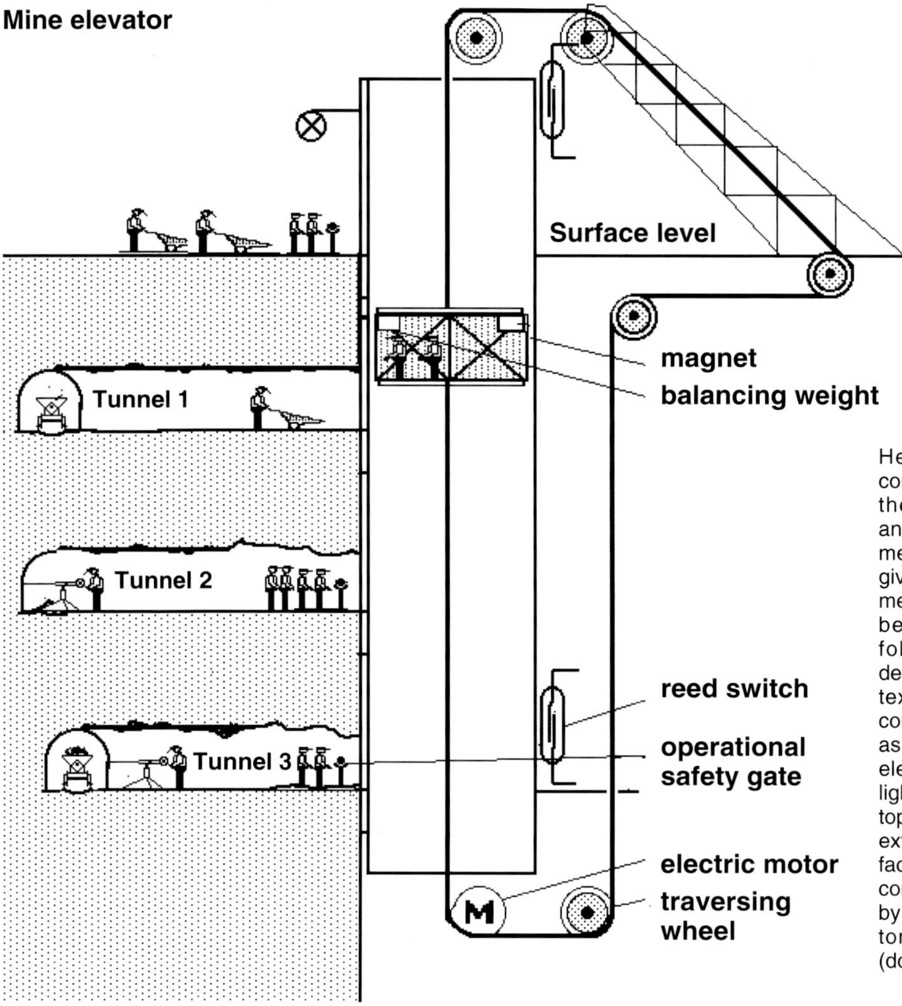

Here is yet another computerized work of art — the artist shall remain anonymous. The elevator mechanics are shown to give you an idea of how mechanical operation can be accomplished. This follows closely the description given in the main text. For the most part, connections are the same as on the construction site elevator. Entry and local lighting corresponds, from top to bottom, to D3-D0 and exterior lighting at the surface level to D4. A bucket conveyer can be controlled by D5 and the elevator motor by D6 (up) and D7 (down).

Mine elevator wiring

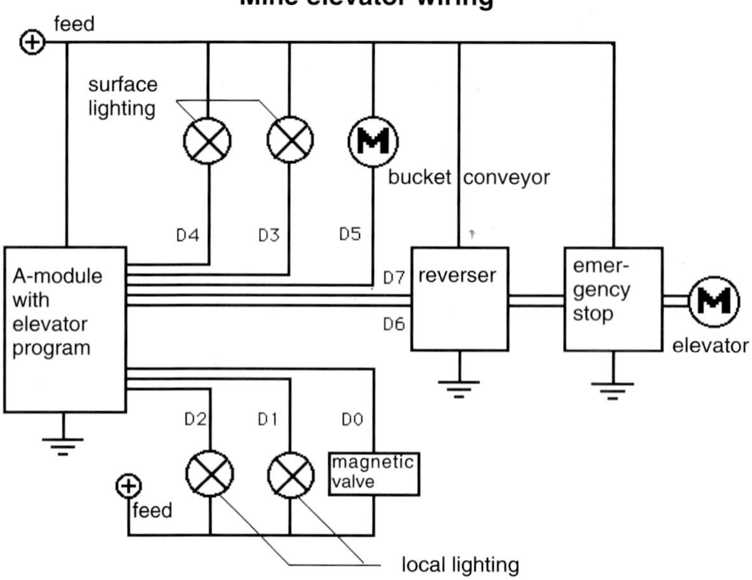

Although this illustration is primarily supplied for those building the mine elevator project, connection details are easily translated for use with the other elevator projects. Note that the emergency stop function is shown in a separate figure.

Ore elevator

The ore elevator differs from the others by having just two stopping positions. The motor is connected as on the mine elevator. An ore bunker is maneuvered with an electromagnetic valve. Operational bunkers are standard with Arnold Rapido's high quality yet reasonably priced, coal works.

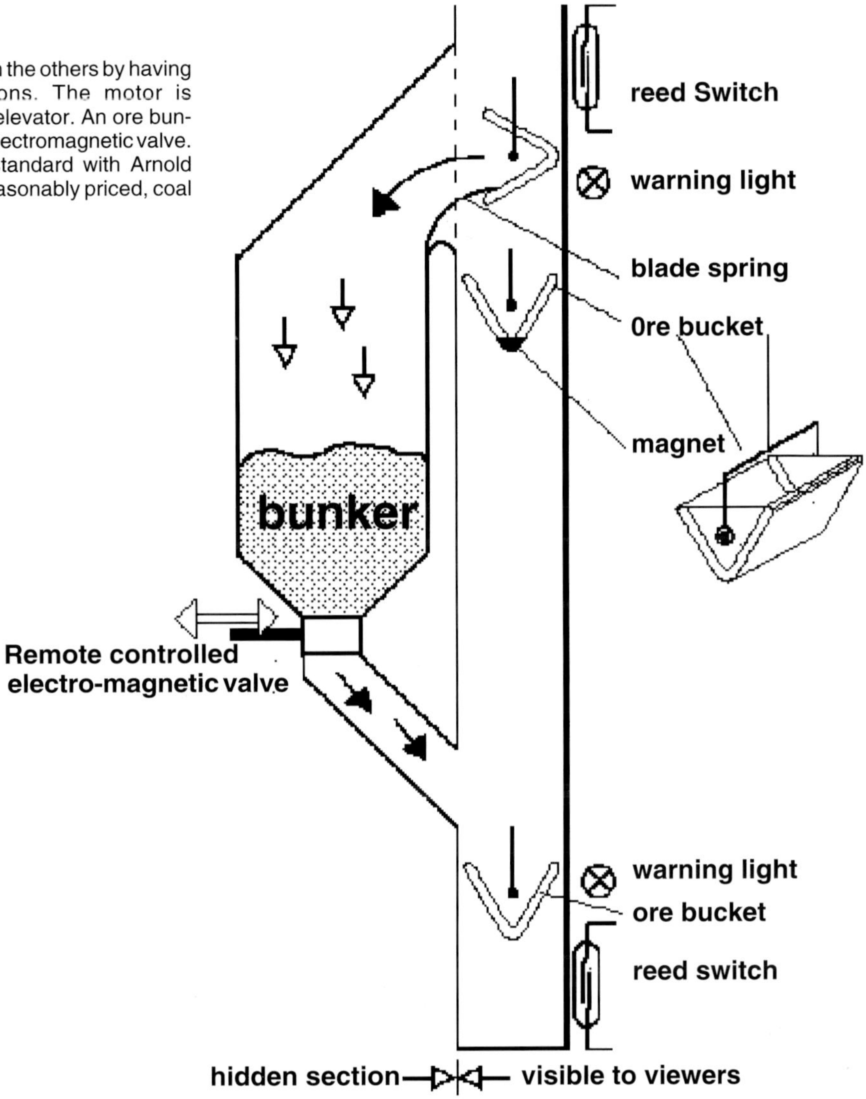

Emergency stop

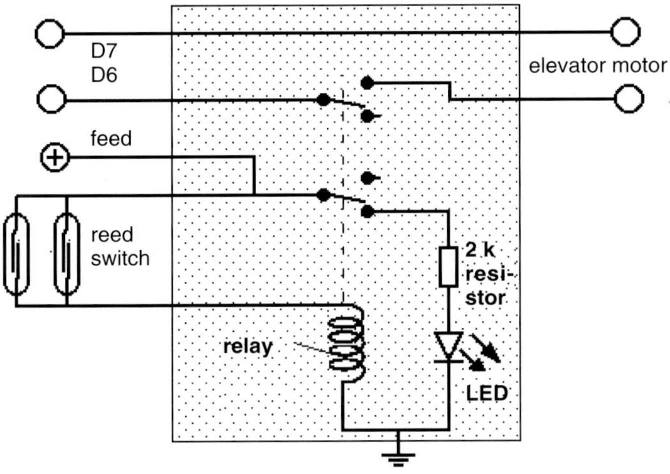

Actually, the emergency stop consists of just a relay with control panel-mounted, warning lights. The reed switches are the same as shown at other positions on the project figures and used to prevent "crash landings". During testing, there is every reason to stay within reach of the power pack's shut down switch.

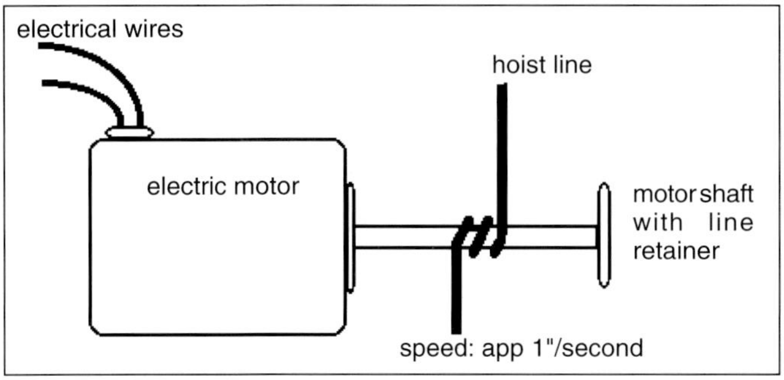

Drive mechanism

All of the demonstration project drive mechanisms have been built in this manner. Depending upon motor speed, the motors have had drums varying from "bare axles" to thread spools from my wife's sewing machine - don't tell her!

EPROM — ELEVATOR, PROGRAM 1

Ore elevator, source code in operational order

12	10	3A	38	1A	18	9A	98
98	98	98	98	98	98	88	88
88	88	88	89	88	81	00	01
04	05	00	01	40	41	40	40
40	40	40	40	40	50	50	58
58	5A	58	5A	18	1A	18	1A
18	1A	18	9A	98	98	98	98
98	98	98	88	88	88	88	88
89	88	81	00	01	04	05	40
41	40	41	40	40	40	40	40
40	40	50	50	5A	58	5A	18
1A	18	1A	18	9A	98	98	98
98	98	98	90	90	80	80	80
80	80	80	80	01	00	01	40
45	44	41	40	41	40	40	40
50	58	58	58	5A	58	52	10

EPROM — ELEVATOR, PROGRAM 2

Construction site elevator, source code in operational order

12	10	3A	38	1A	18	9A	98
98	98	98	98	98	98	88	88
88	88	88	89	88	81	00	01
04	05	00	01	40	41	40	40
40	40	40	40	40	50	50	58
58	5A	58	5A	18	1A	18	1A
18	1A	18	9A	98	98	98	98
98	98	98	88	88	88	88	88
89	88	81	00	01	04	05	40
41	40	41	40	40	40	40	40
40	40	50	50	5A	58	5A	18
1A	18	1A	18	9A	98	98	98
98	98	98	90	90	80	80	80
80	80	80	80	01	00	01	40
45	44	41	40	41	40	40	40
50	58	58	58	5A	58	52	10

EPROM — ELEVATOR, PROGRAM 3

Mine elevator for passenger, equipment and ore transport, source code in operational order

12	10	3A	38	1A	18	9A	98
98	98	98	98	98	98	88	88
88	88	88	89	88	81	00	01
04	05	00	01	40	41	40	40
40	40	40	40	40	50	50	58
58	5A	58	5A	18	1A	18	1A
18	1A	18	9A	98	98	98	98
98	98	98	88	88	88	88	88
89	88	81	00	01	04	05	40
41	40	41	40	40	40	40	40
40	40	50	50	5A	58	5A	18
1A	18	1A	18	9A	98	98	98
98	98	98	90	90	80	80	80
80	80	80	80	01	00	01	40
45	44	41	40	41	40	40	40
50	58	58	58	5A	58	52	10

This is Europe meeting US in a model rendezvous. The Big Boy and the Swedish heavy class A engine both require some organised coal handling operation. Our Iron ore project can be turned into coal supply where the bottom of the looped storage system then is opened up to load the tender. I you arrange a small hole vertically in the tender the "coal" will slowly disappear onto the track and ideally a track feeding system below.

MINI-TIP

A handheld controller

This project was based upon the common, household lighting dimmer. In this version of the circuit, thyristor control is adapted for the low voltage available at a power pack's AC lighting outputs.

The throttle is built around a thyristor of either type C106 or C122. The AC voltage is converted in the full wave rectifier to a pulsed DC voltage. This voltage shouldn't be evened out by a capacitor. Instead, the thyristor sends these pulses to the locomotive in pulse widths proportional to desired engine speed. This is regulated by a potentiometer. Because DC locomotive direction of travel is determined by track polarity, a polarity switch is included in the controller.

With two controllers and the addition of a second engine, the beginning model railroader can increase operating complexity. The layout can be divided into a mainline block and a yard block.

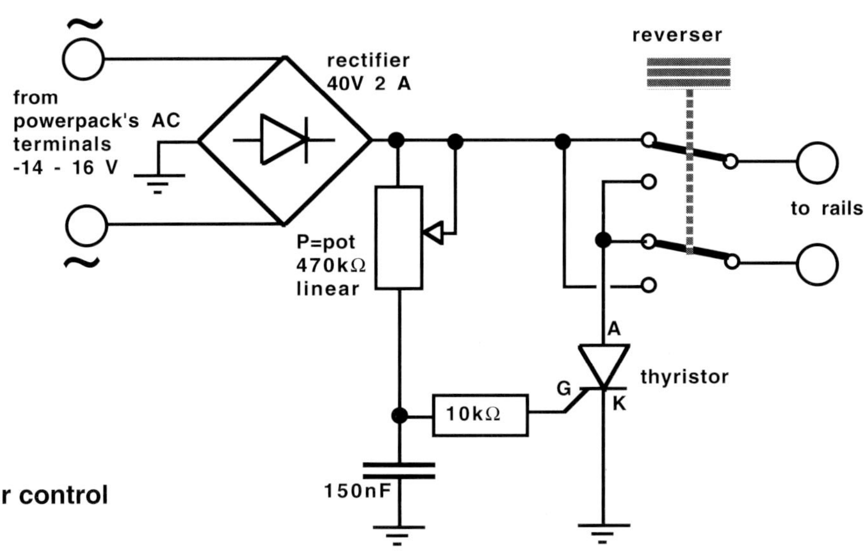

Thyristor control

Programmed module:
Life at the fire station

This section takes up a typical 24 hour day at a fire station. Fire stations and fire engines are quite common on model railroads but because nothing happens - nothing is heard, lit or flashed - attention is not directed to them. A viewer can miss them entirely.

We are going change that. Come along on a construction tour of all thinkable electronic modules that can liven up the old fire house. We'll add sound, flashing warning lights and general lighting. The fire house doors will open and close and it will all happen in a compressed 24 hour period. Let's start our tour.

With rather inexpensive and uncomplicated electronic components we are going to put together a sound and light show. There is something magic about fire stations. After railroad stations and platforms, fire stations are probably among the most popular structures on a model railroad layout. The wide assortment of fire-fighting vehicles and figures available certainly contributes to their popularity.

I previously described how to simulate fire with smoke and flames (electronic flickering light, that is). Visitor interest in my electronic fires inspired me to do something about my fire station.

THE MODULAR SYSTEM PROVIDES FLEXIBILITY

In earlier projects, I have used modular design with a certain amount of success. The modular design philosophy fits well here. It's like a restaurant menu - the menu offers several different dishes but you only order the one you want. You can start with a small project and later expand upon it if the first stage goes well. Perhaps you would like to concentrate your efforts on sound and movement in that the station is already bathed in light. Or maybe the structure isn't a fire station but an apartment house, an office building or small industry with several lighting points to be controlled. From here on, we will focus on technique, leaving application to the reader.

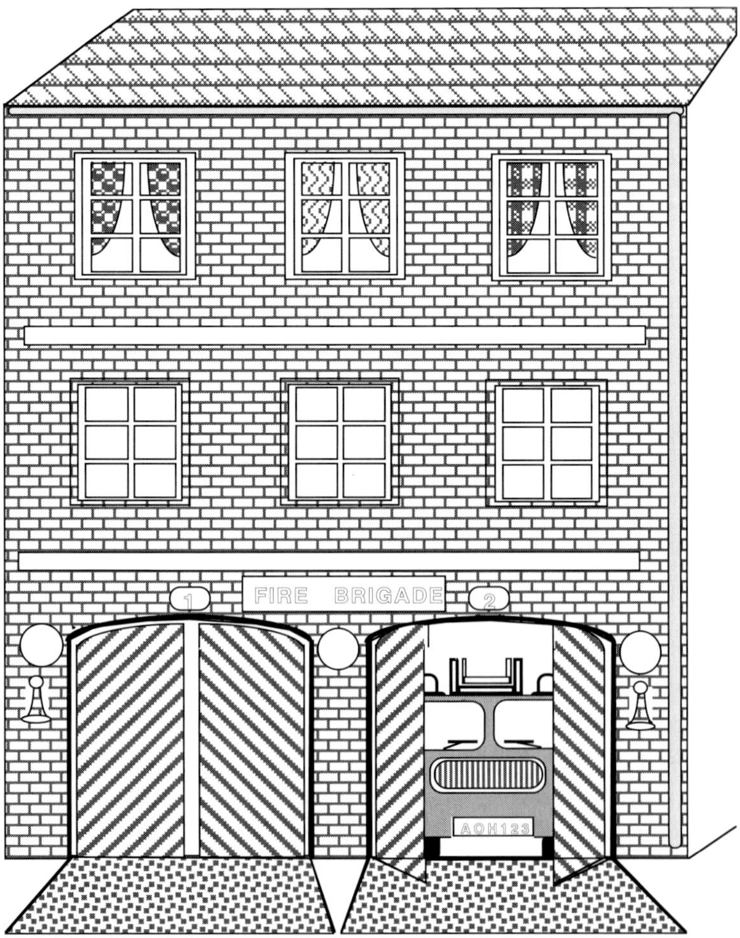

C-module (central selector)

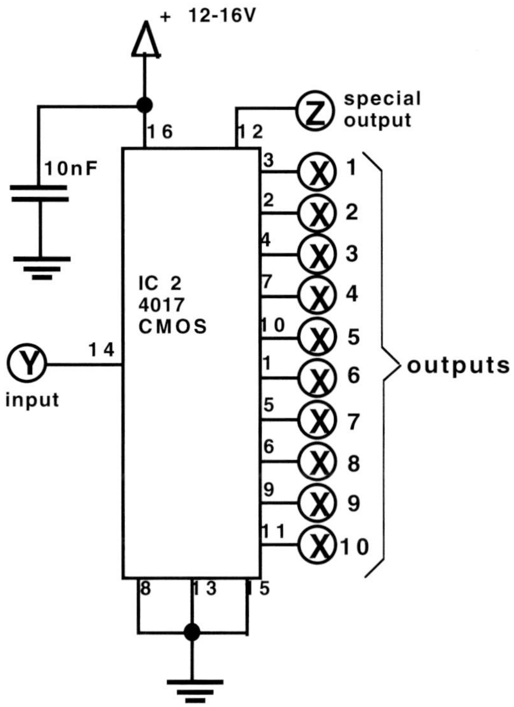

IC2 is the central event selector. All outputs are normally low (0V). Only one output at a time is high (meaning equal to the power feed voltage of 12-16V). The IC is called a "walking ring" and climbs continuously, round after round - after output 10, comes output 1, etc. Output Z can be used for longer events. It goes high after steps 1-5 and low for the rest of the "round". ICs can be cascaded to attain additional control steps. Cascade connection is not within the scope of this project.

L-module (lights)

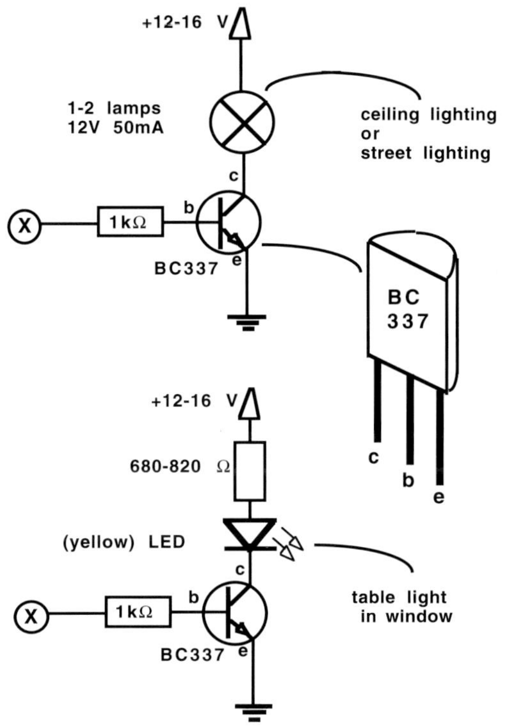

The lighting module is in two versions - one for lamps and one for LEDs. As long as just one or two points of light are controlled, a BC337 is satisfactory. For controlling considerably larger loads, a 92PU45 transistor is recommended. This transistor is used on the R-module.

T-module (time control)

The length of the station's "24 hour day" is controlled by this central clock. In it's basic configuration, it divides the day into ten parts. Refer to the capacitor/ time chart below to set the day cycle length by varying the value of C1. For system stabilization reasons, the capacitor at pin 5 is vital and must not be excluded.

Values for C1 - electrolyte capacitor:

1µF ->1 cycle every 30 seconds (approximate)

10µF ->1 cycle every 5 minutes (approximate)

100µF ->1 cycle every 50 minutes (approximate)

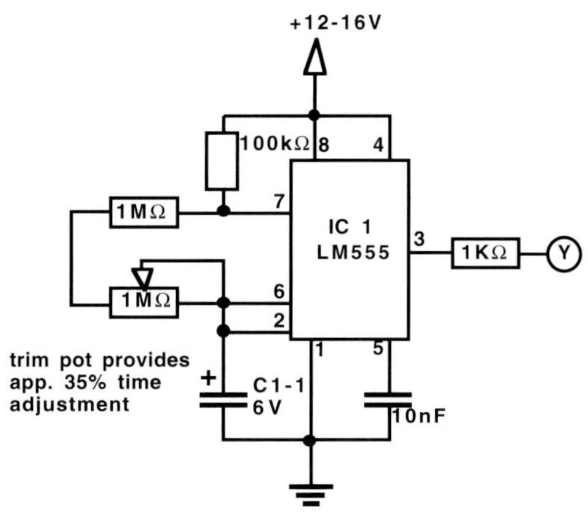

Special diode circuitry for combined outputs

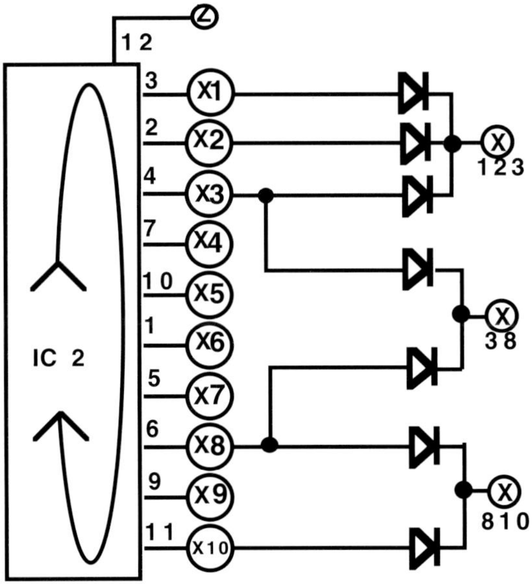

All IC2 outputs are at low logic level except the one momentarily active (high logic level). The Z output is high when outputs 1-5 are active and low when outputs 6-10 are active.

Thanks to blocking diodes we can add new, output control combinations to IC2. Output X123 combines the active time from the first three outputs. That means that a lamp connected to output X123 will shine continuously over three periods.
In a corresponding manner, X810 is connected to outputs 8 and 10 and will be active for two periods. X38 illustrates another control principle. Here, output pins 3 and 8 are connected to two separate sub-circuits. When connecting outputs to multiple sub-circuits, capacity limitations must be considered. As long as we are driving control operations via a power transistor, three or four pins may be safely connected in this manner.

B-Module (flashing lights)

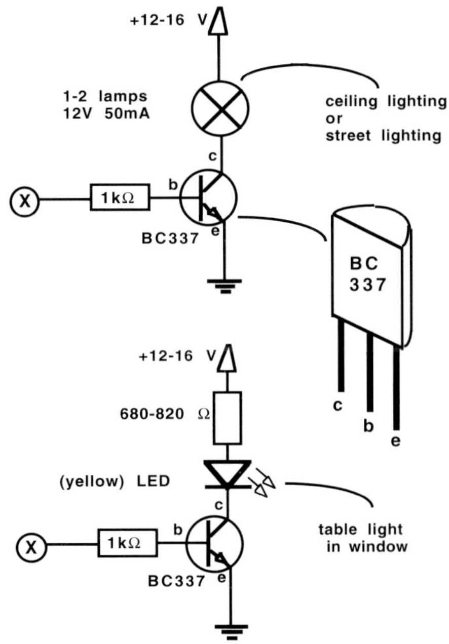

Flashing lights can be used at several locations. Red flashing lights warn motorists as fire trucks roll out from the station. Warning lights are also switched on when opening fire house doors. These are often orange or red. Blue LEDs are also available but are expensive, at present. These warning light units both have one common characteristic — the warning lamps usually flash alternately.
Note that the module's voltage supply is via the X connection. If several B-modules are run simultaneously, power transistor control should be incorporated. Here, a CMOS version of the 555 timer circuit is used. Its current drain is less than the common version but it is slightly more expensive. Although I have used common LM555 ICs in the past, they should be avoided - they put too much strain on IC2.

USE READY-MADE PRINTED CIRCUIT BOARDS

Small modules of this type are quick and easy to assemble on ready-made circuit boards. If you haven't seen these before, they are pre-drilled boards with flat, parallel copper leads on the soldering side. The board is peppered with holes spaced at 0.1" intervals. That means that all common components such as ICs, relays and transistors can be set directly in place and soldered without drilling new holes. The copper leads are used for wiring. They can be cut with a hobby knife and interconnected with wire jumpers. These jumpers are usually mounted on the component side. Often, long component leads (especially on capacitors and resistors) can serve as jumpers. My circuit boards follow the European standard and measure approximately 4" x 6" and 4" x 12". An American standard circuit board, corresponding to the larger format is big enough for the entire project. You can collect all modules on one board (this will give you a good circuit overview and minimize leads) or divide the board into smaller subsections.

Sectioning is done by scribing both sides of the board with a knife, from hole to hole. The board can then be snapped in two. It is important that the copper leads crossing the division are cleanly cut before snapping.

A common construction method is to begin component attachment from one of the ends. When the module is finished, you can see how much space was required and shorten the board accordingly. A practical assembly guideline is to place components so that they are easily replaceable. It shouldn't be necessary to remove five good components to get at one bad one. You may sometimes be tempted to even mount components on the soldering side. Consider this alternative as a last ditch solution. Soldering side components are easily moved and either the component

S-Module (station alarm)
One-tone siren

The S-module is based on a sound circuit from the German manufacturer, Siemens. The SAE0700 is small, economical and requires no external sound amplification. Connected as shown, it produces a long-duration tone. Tone pitch is adjusted with the trim potentiometer. The speaker can be hidden from view under the layout or in the fire station.

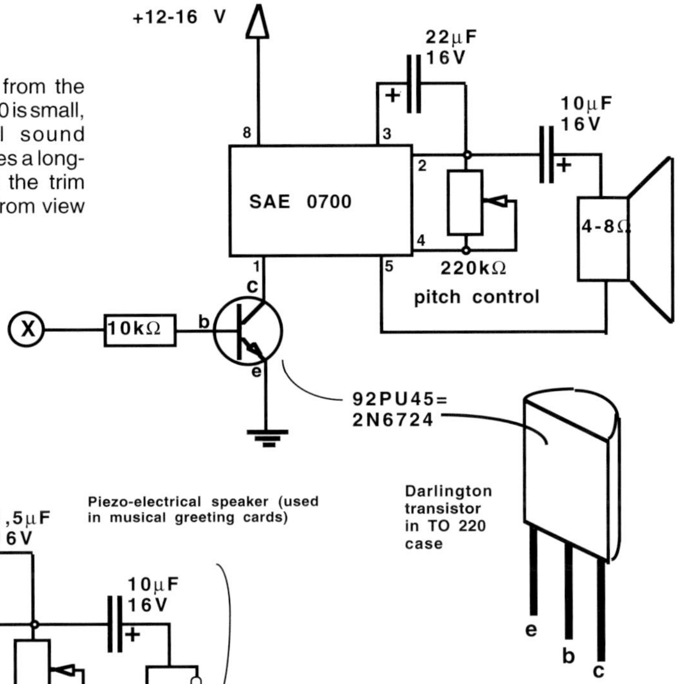

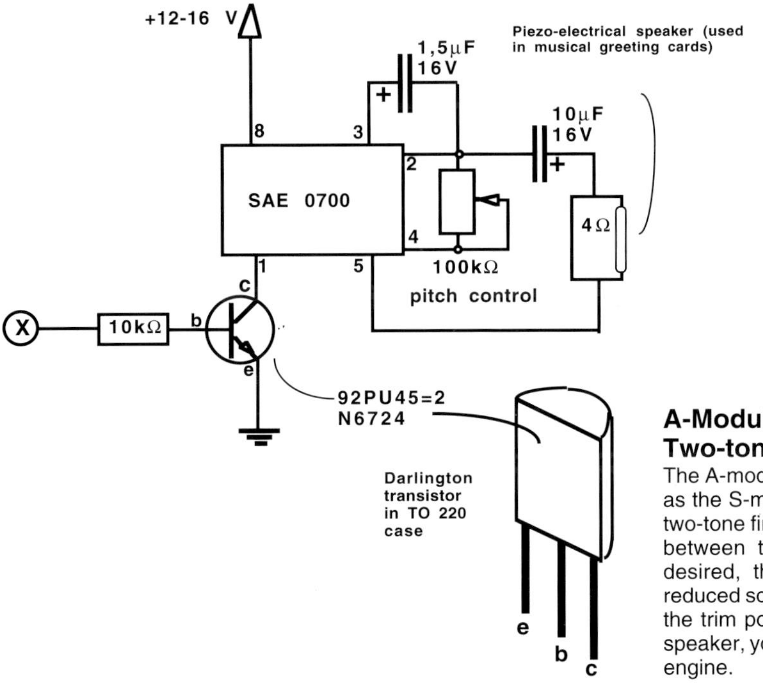

A-Module (siren)
Two-tone siren

The A-module uses the same integrated circuit as the S-module. This time, we will simulate a two-tone fire engine siren. The siren will switch between two tones. If quicker switching is desired, the 1.5µF capacitor value can be reduced somewhat. Tone pitch is adjusted with the trim potentiometer. If you use a miniature speaker, you can probably hide it in a model fire engine.

itself or its leads, can short circuit on the board's copper leads.

SMART PROGRAMMING WITH DIODES

The central component is a 4017 CMOS counter IC with an aversion to static electricity - it nearly goes without saying that you don't have wall-to-wall carpeting in your hobby corner. The counter is of the "go around" type and steps through all ten outputs with one step per pulse from the T-module. After the last output, it starts from the beginning in a continuous ring-cycle. Programming in the basic version is a bit primitive when each output is just as active as its neighbor. Each sound or lighting effect is switched on one at a time and in the same order - never in combination with one another. With the help of diode type 1N4148, we can achieve interesting output combinations in accordance with our own needs and tastes.

To produce more than spectacular nonsense shows we should first prepare a time flow chart - a chart indicating events at the fire station in relation to time (when activities should be started and stopped). A suggestion for just such a chart is provided at the end of this section. With diodes, the correct event combinations can be produced. Hopefully, the subject-specific descriptions in the figure captions are sufficient for building the project,as opposed to the more general information given in the main body of text.

R-Module (relay for door opener)

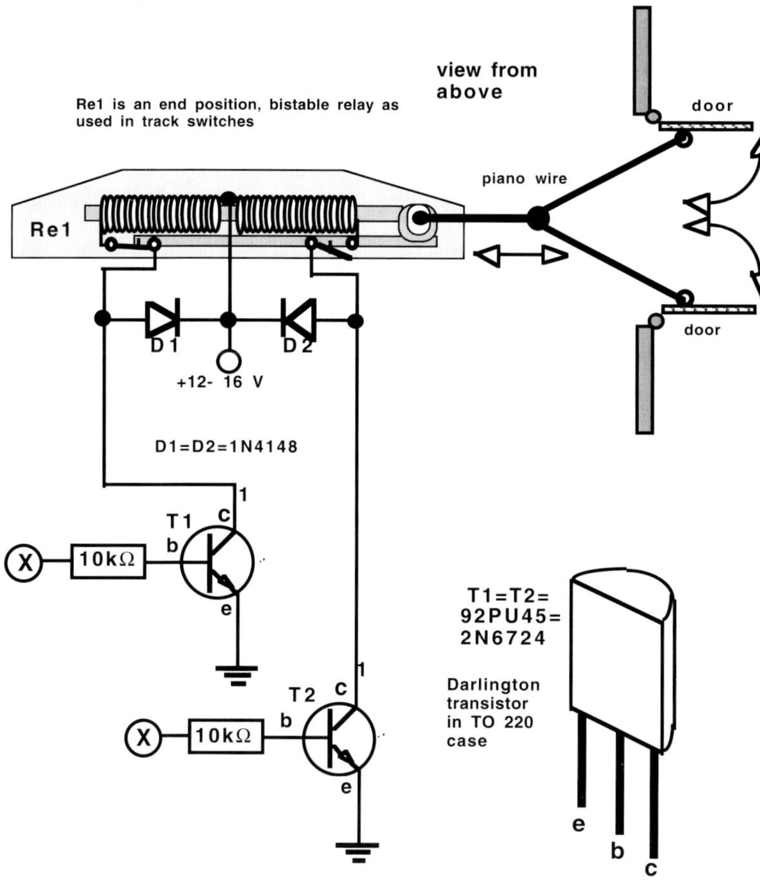

This module requires a little more handiwork over and above the electrical skills used earlier. It can be worth the effort — all movement on a layout tends to draw attention. Half the job is done if you use a track switch, electro-magnet with an end position switch. The action is just right for opening and closing doors. If slower opening and closing is desired, a switch motor can be used instead. The entire construction is hidden from view under the fire house roof.

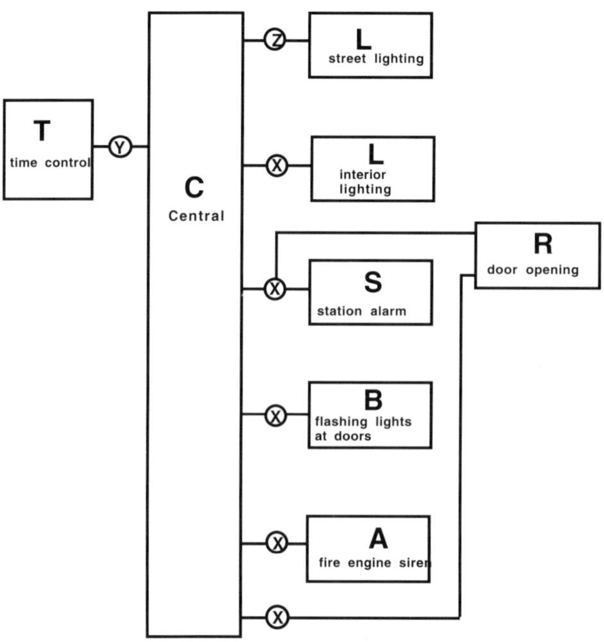

How to connect the system

This is an overview of the entire system. There is nothing to keep you from using several modules of the same type to maximize station activity. If a second Central-module is cascaded with the first, an expansion can be used for the “second half” - a sound and light show at a burning building. If this sounds like a good idea, use a second set of fire-fighting vehicles and shut down fire station operations as fire site effects are switched on for increased realism.

This is the almost complete command control unit for the Fire Brigade Station. It includes the R and S modules for dooropening and sound alarm. It is often very time saving to wirewrap or lab-connect an electronic module before soldering the components finally. Individual components for leadtimes and sound frequencies are then easy to change.

This is the suggested time code for station events. It starts when the station alarm sounds and street lighting is switched on. A moment later, interior lighting is turned on and the doors open. The warning signals at the street begin to blink, the fire engine's lights start flashing and the siren is turned on. A while later, the doors close and all is peaceful until the next cycle begins. Using this table, appropriate combination, diode circuitry (see the separate figure) can be used. The station alarm could, for example, be combination-wired to output X123.

EVENT TIME CODE

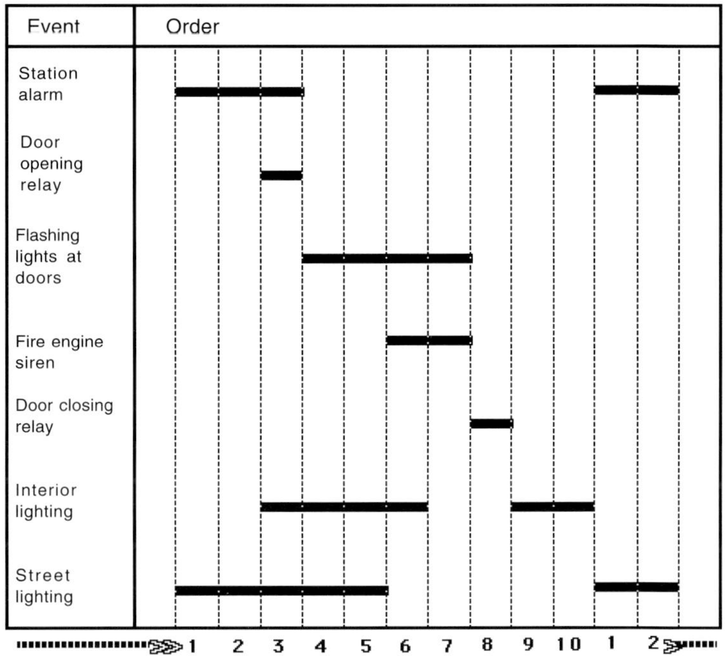

outputs at IC2 and combination circuitry

Sirens for emergency vehicles

New sound systems are always welcome. A vehicle still life at the scene of a fire attains completely different dynamics and vitality when complemented with flashing lights, smoke and assorted sounds. In this section, we'll examine several different warning sounds based upon a common electronic building block.

Warning sounds are certainly a whole science in themselves. Here we will limit ourselves to mirroring reality. Sound must be experience and cannot be difficult to describe in print.

With this circuit we can create more traditional, audio warning signals based on European prototypes. The sound of the traditional, American siren is not reproducible with this unit.

ONE IC DOES HALF THE JOB

The German company Siemens has introduced a smart IC containing double tone generators. It is designated SAE0700 and is the foundation of this project. Sound pitch is regulated with external resistance rated at 1.5kΩ to 300kΩ. It provides a frequency register that corresponds to all nearly audible sound except for the bass range (100Hz to 15kHz).

Switching intensity between high and low tones is regulated by an external capacitor rated in the 15nF to 1.5µF range. At one extreme, switching occurs every other second and at the other extreme, 50 times per second.

The circuit drive a small, piezo-electric buzzer (used in musical greeting cards) or high resistant, miniature speakers.

POWER FEED: "USE WHAT YOU HAVE."

Generally speaking, the circuit can be powered by the power source of your choice. Full polarity protection is present between leads 1 and 8. This means that plus and minus from a DC source can be connected in any manner. This also means that the IC can be connected to an AC source in the 10V to 28V range (and even up to household voltage levels for up to 30 seconds for other applications). DC power sources should be in the 10V to 20V range.

If you have only a 9V source available, connect minus to lead 2 and plus to lead 7. Note that these two inputs are not polarity protected.

IC function is such that tone generation starts at power-up and continues until current is interrupted. Sound reproduction always begins with the higher tone.

SOME PRACTICAL ADVICE ON ASSEMBLY

The unit is so compact that it can be easily installed in an HO scale vehicle. It can also find room in larger trucks and busses in N scale. With Z scale, you'll have to tuck the unit under the layout or hide it in another suitable location. Be sure that the opening for sound faces upward.

If the vehicle is mounted on the layout surface, run thin

This is an experimental setup for checking circuit limit values. A push-button cut-off switch isn't necessary but practical - a continual siren howl can be unnerving. A homemade handle (not shown) for the trim potentiometers makes for quick and easy adjustments when searching for just the right sound.

leads to the unit's power inputs. An entirely new generation of HO scale vehicles has been introduced by Faller. These vehicles, equipped with rechargeable batteries, can be started, stopped and steered remotely. With these in operation, one could even imagine an interesting train layout with trains. For our needs, voltage for the sound unit must be taken from batteries. Three volts is available onboard. An additional six volts is needed to reach the nine volt level required by the sound unit. Four, 1.5 miniature batteries solve this problem handily.

Suitable starter external components for testing are a 1µF/16V tantalum capacitor and a 50kΩ resistor (subminiature 100kΩ, trim potentiometer is also a good choice, offering greater variations in pitch).

Lastly, a little tip for those cases when available installation space is at a premium. Solder the capacitor and resistor directly to the IC leads and use speakers from a compact headphone set.

Siren for emergency vehicles

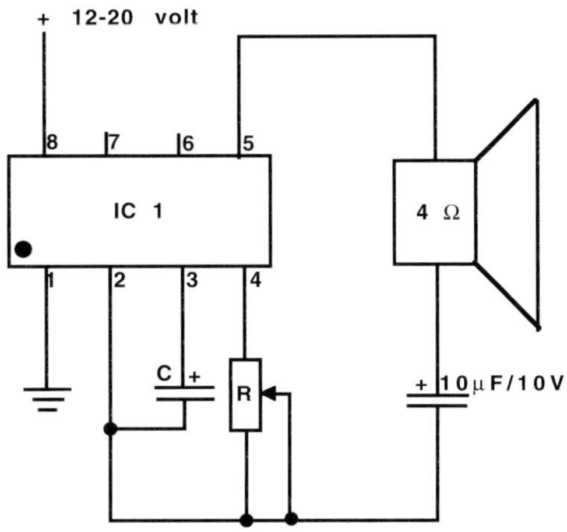

The schematic is quite simple in that most functions are on the ICs internal chip (silicon chip). Besides the IC and a speaker, just one resistor and two capacitors are required. Power feed is shown as a DC voltage with plus to pin 8 and minus to pin 1. See the text for alternative power feeds. Note that capacitor C and resistor R were selected using the formulas provided.

How to calculate tones and switching frequency

Higher tone frequency FA

$$FA = \frac{27200}{R\ (k\Omega)}\ Hz$$

Lower tone frequency FB

$$FB = 0{,}725 \times FA\ \ Hz$$

Switching frequency V

$$V = \frac{750}{C\ (nF)}\ Hz$$

These formulas are provided for calculating component values. If we choose a 27kΩ resistor, the circuit produces a high pitched tone of just over 1kHz in accordance with the first formula. The second formula shows that the low tone will be at just over 700Hz. In the third formula, capacitor C determines tone switching speed. With a value of 750nF (that's equal to 0.75µF) we'll get one switching per second, or 1Hz.

Siren limit values

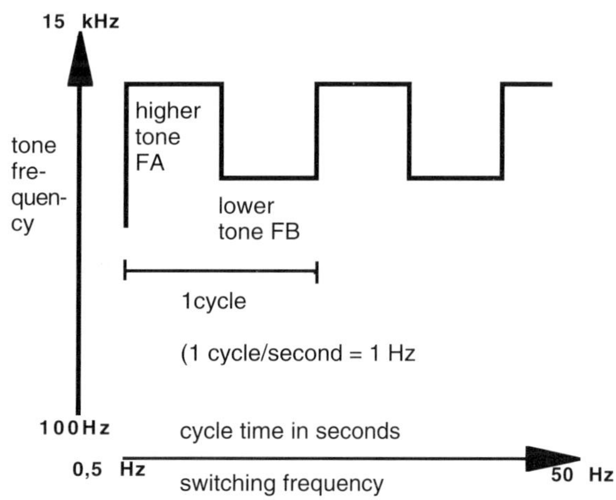

The Siemens circuit is universal and has a very broad audible spectrum, far wider than what we need here. The module's single resistor can be vary the two siren tones from 0.5Hz (1 switching/2 seconds) to 50Hz (50 switchings/second). At 50Hz, the sound is difficult to discern other than as a blended tone. Note that the letter designations are the same as used in the formulas.

More program module experiments:

The smart grade crossing for all scales and systems

A full-fledged grade crossing is an excellent object for control by our program module. There are several control points - gates, red flashing warning lights and even bells. A grade crossing characteristic is that all functions repeat for an infinite number of times and without variation. A program segment in the termination stages (crossing gates on the way up) should also be possible to break if a new train is on the way.

The system is completely isolated from train control. For this reason, it is well suited for all systems and scales. The power pack can supply the operating voltage as long as it is at least 6.5V The internal voltage stabilizer sends 5V out to the ICs. The remaining 1.5V is required for conversion operations. The base module, A, converts AC to DC permitting direct connection to AC power pack terminals.

Swedish grade crossings have a white, flashing all-clear signal for vehicular traffic mounted under the red signal lamps. This description includes this feature.

A RATHER SINGLE-MINDED PROJECT

The first grade crossing variation is useful for single track with one-way traffic. It requires relatively little in the way of control electronics. A schematic of the circuit is provided. We'll need both modules A and B. A chart showing module connection is also provided. The entire system can be magnetically controlled with small, powerful magnets glued to the locomotive undercarriages. When a train approaches the crossing, the track-mounted reed switch is tripped and the high voltage level at P12 is momentarily grounded. This starts the program we'll soon be taking a little closer look at. Just beyond the crossing, a second reed switch triggers the deactivation subprogram that raises the gates and returns the program to the standby state. You can follow this sequence by using the two, adjoining program charts. You will see exactly how signal lights, gates and warning bells are controlled by the train and how the program can automatically jump from the deactivation to

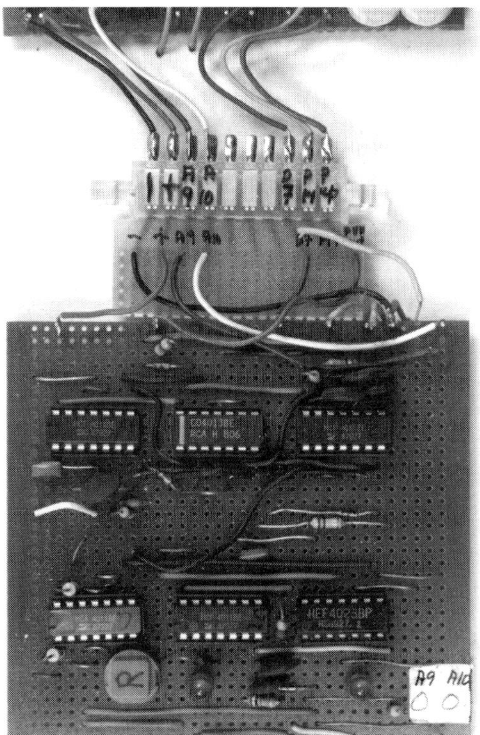

The B- and C-modules — photo of a larger circuit board with three leads with connector equipped leads. The combined B/C unit has plenty of space for components, making it easy to build. The connections are to the power pack.

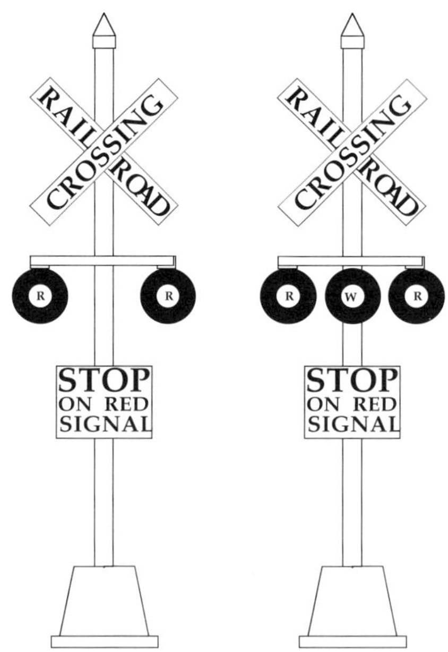

Both old and modern railroad crossing signposts use red flashing lights to indicate that train is coming. These red lights may in N-gauge and H0 be LED´s that can be connected directly to the output of the module A.

If you want to indicate "permission to cross the track" the internationally accepted sign is slowly and soft flashing white light. A compromise is to use a yellow LED, but if you have a small bulb available there is a special module in this chapter for ramped flashing white light, that fits directly to the D0-output of the A-module.

Motor module

Although the module is just 3/4" x 1 1/4" in size, it contains the entire motor servo, including its four power transistors. If you have plenty of space under your layout, you naturally do not need to build as compactly.

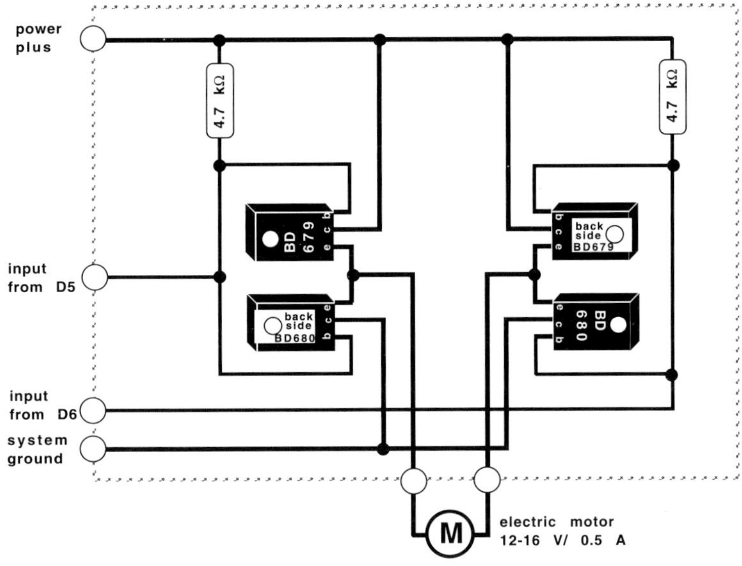

the new-start program segments via D7. D7's sole purpose is to initiate new starts.

The B module is shown in a separate schematic and consists primarily of two RS flip flops - either 4011s or 4014s. For more information, see the chapter covering lighting in houses and apartment buildings.

TWICE AS MUCH - TWICE AS GOOD!

This showpiece grade crossing version is for bi-directional operation and with double track. The circuit is capable of dealing with one more than one train in the crossing trackage. If another train approaches the crossing as the gates are on the way up, the new-start subprogram is reactivated. Reed switch connection is a bit different from the single track version and we will also need a combination of A, B and C modules. A photo and schematic show my prototype with the B and C modules on the same board. There are two LEDs at A9 and A10 (address inputs). As you can see in the program chart, these LEDs indicate which of the subprograms are active.

Using the connection chart and the overview diagram showing all complementing connections, you can easily follow the signal path from reed switch to control output.

SERVO CIRCUITS

During my experimentation with program control, a number of small, practical circuits were put together, designed exclusively for use with grade crossings. I have summarized these in separate figures with accompanying, descriptive text. One circuit that was especially effective simulates the Swedish prototype, all-clear signal, mentioned previously. It takes about two seconds for this signal to go from off to on, and off again. Light intensity rises and falls successively. This is in direct contrast to the clipped light from a lighthouse, for example. Adding this soft flash feature calls for just one extra transistor and a few discrete components.

This is a AC driven bell from Brawa. As long as the original sound is single tone that works perfectly with the project programme. This is also the most common in the US. But in countries like Britain and Sweden you hear dualtone bells at road x-ing and therefor the program output allows both versions.

You will find the two outputs D3 and D4 on the command output table which is close to this text and they are assigned for separate bells.

Single track with one way traffic

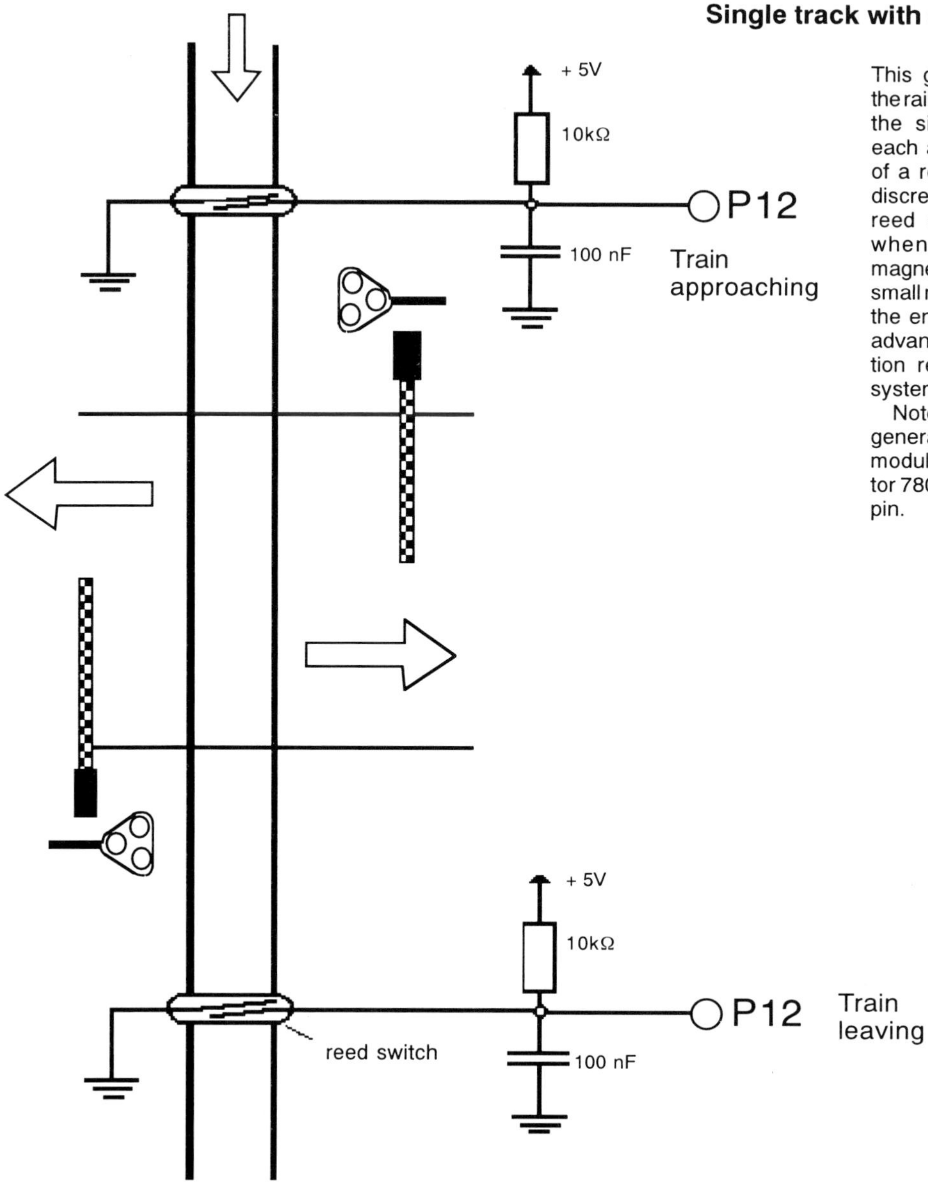

This general overview of the railroad crossing shows the simple assembly of each activator. It consists of a reed switch and two discrete components. The reed switch is activated when brought into a magnetic field. Here we use small magnetos glued onto the engine which has the advantage of safe operation regardless of power system or gauge .

Note that the +5volts is generated from the basic module where the regulator 7805 has a 5volt output pin.

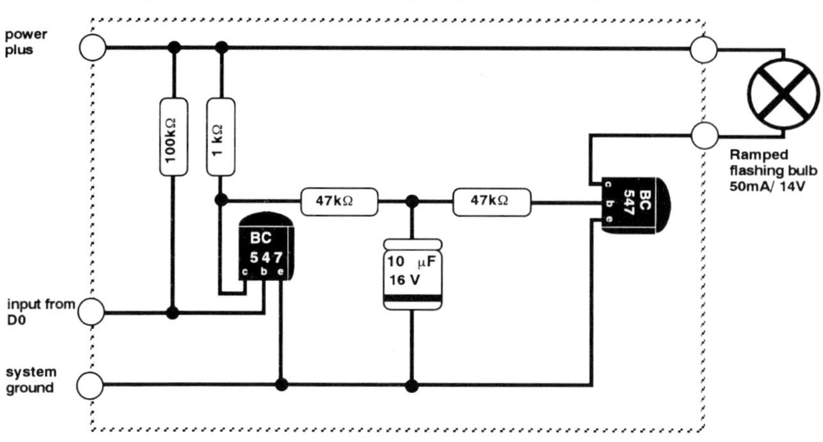

Module for ramped flashing

Soft-light module — photo of a long and narrow electronic module on a circuit board with five leads.

This insignificant looking module is used for control of the "soft" white warning light as used on Swedish grade crossing signals. The schematic is provided in a separate figure.

Double track with bi-directional traffic

Demands on our electronics increase...The trains can come from both directions and at unsuitable occasions, such as when another train is already in the crossing block. By combining functions from modules A, B and C, the system can handle these difficult situations as witnessed by extensive testing on the author's layout.

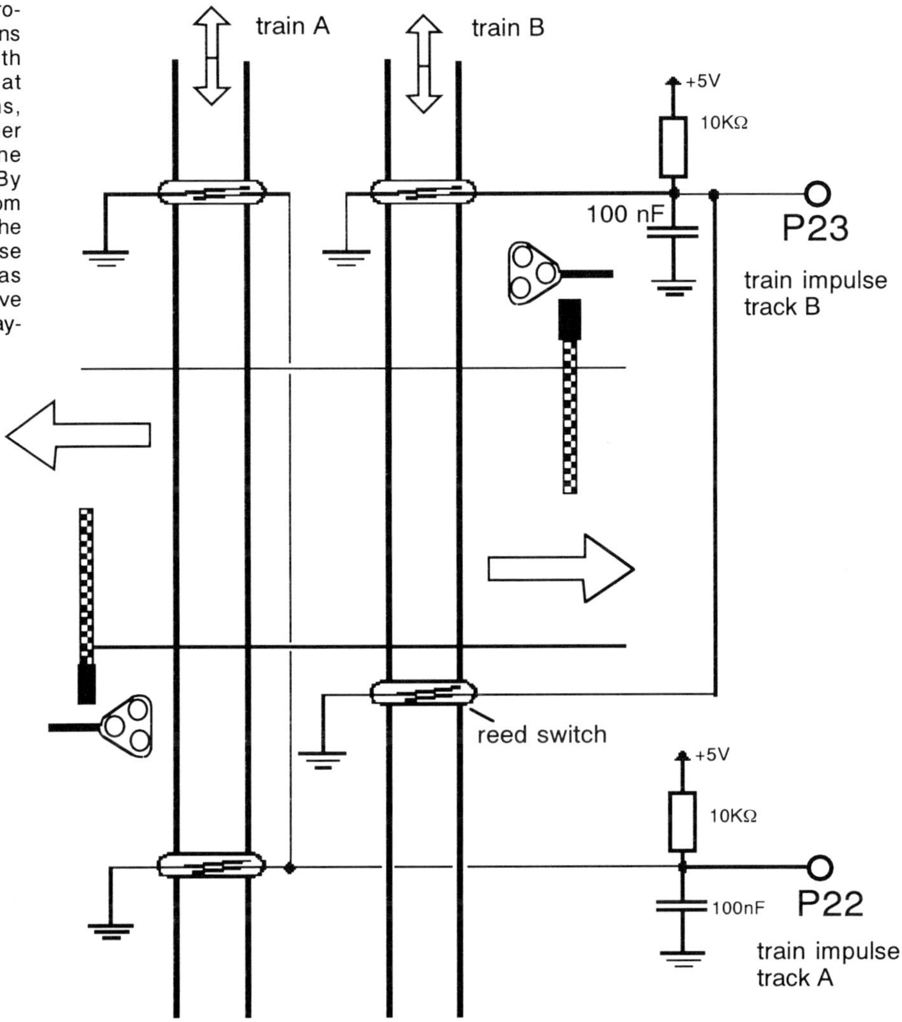

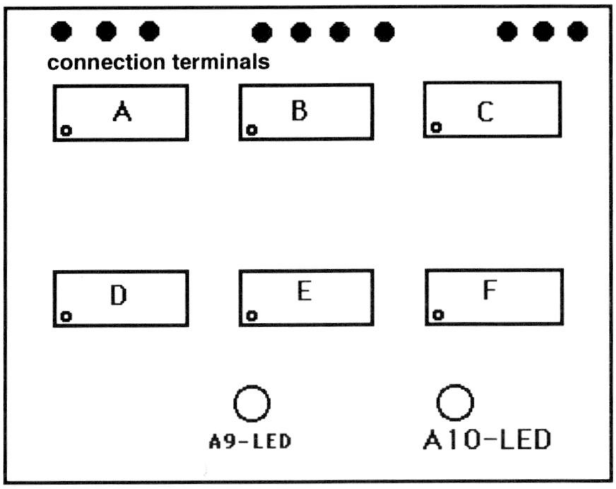

B/C module combination

Prototype component-layout

The B and C modules can be easily combined on the same circuit board for double track applications. This figure shows the component layout pictured in the photograph of the author's prototype. Note that the circuit has been complemented with LED indication for both address outputs A9 and A10 to permit complete program identification. For normal use, it's enough with A10. The LEDs are not needed on additional B/C module units. Note that these LEDs have nothing to do with model operation.

Program table for grade crossing

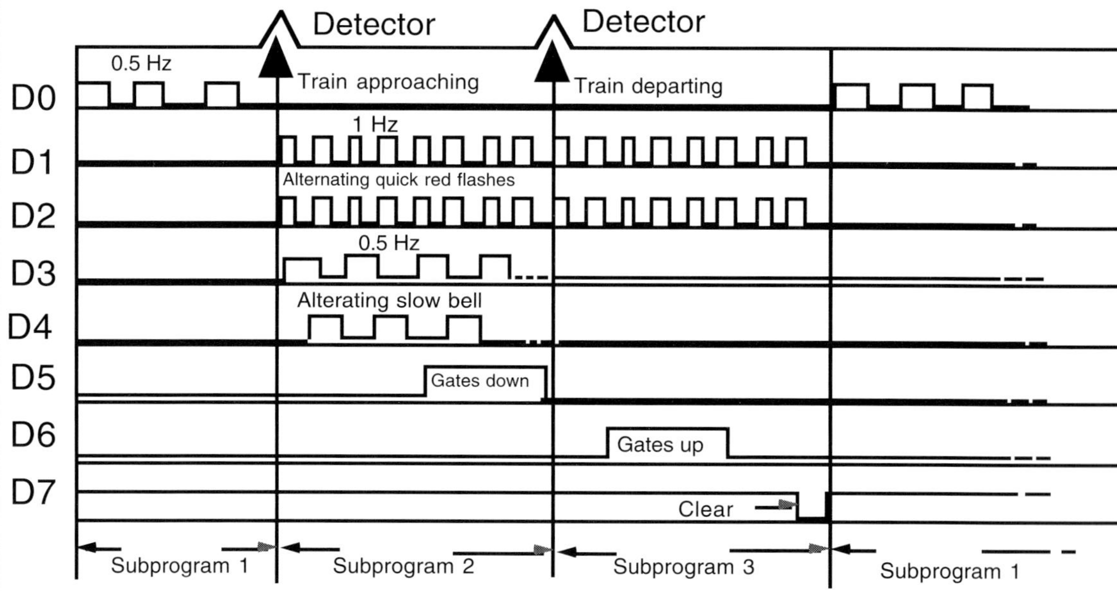

Program flow table

FLOW	A10	A9	EPROM ADDRESS	SUB-program	Pgm No.
□	0	0	0-511	CLEAR TRACK	1
	0	1	512-1023	TRAIN approaching	2
	1	0	1024-1535	Not used	
	1	1	1536-2047	TRAIN departing	3

□ = **Starting point at power up**

The upper table illustrates how output signals are coordinated with one another in the various subprograms. With the program flow table, you can easily follow how trains steer subprogram selection. Subprogram 1 is designed to run continuously until subprogram 2 is activated. As with the other subprograms, subprogram 2 runs at 1 Hz and has 512 program steps. This means that subprogram 2 has a duration of several minutes. This is nearly good enough for prototype applications and certainly sufficient for model railroad use.
Termination subprogram 3 is activated as soon as the train leaves the detection block and lasts but a few seconds. Note that even if termination has begun, the program restarts if another train enters the block.

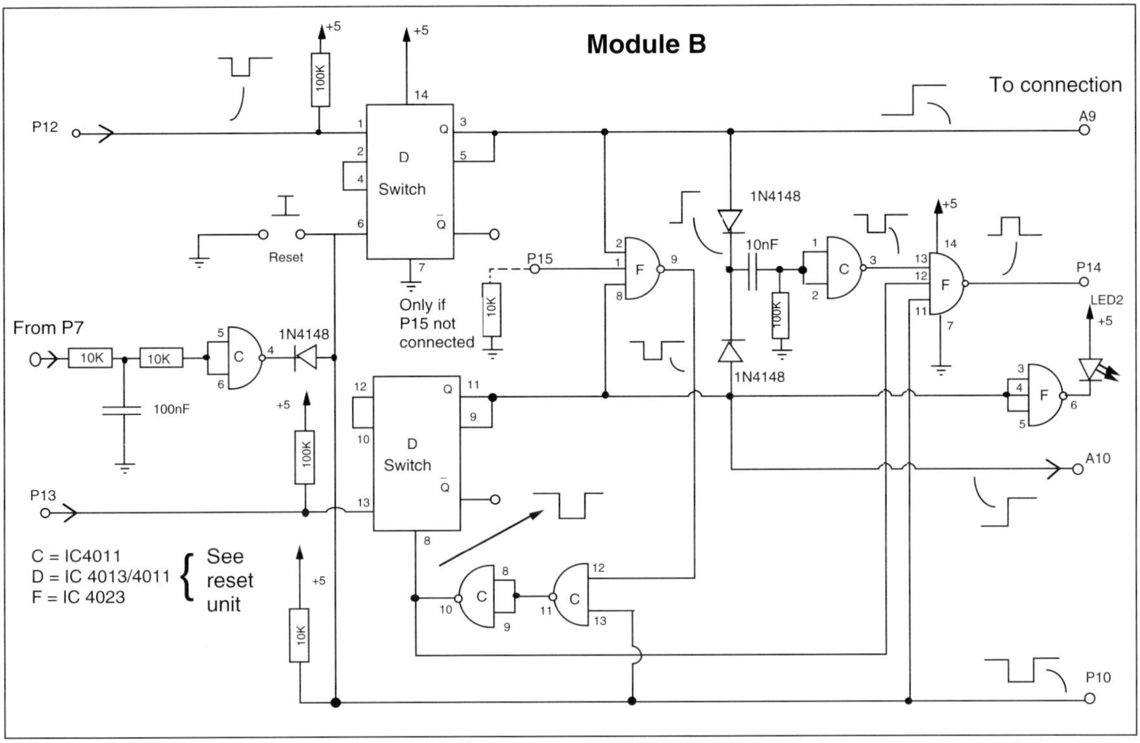

MODULE B — SCHEMATIC
The simpler grade crossing version for single track applications requires module B only. There are two RS switches, which can be constructed using ICs. See the schematic for the structure lighting program. Note that both A9 and A10 can have LED indicators as long as buffers (such as inverters) are used as LED drivers.

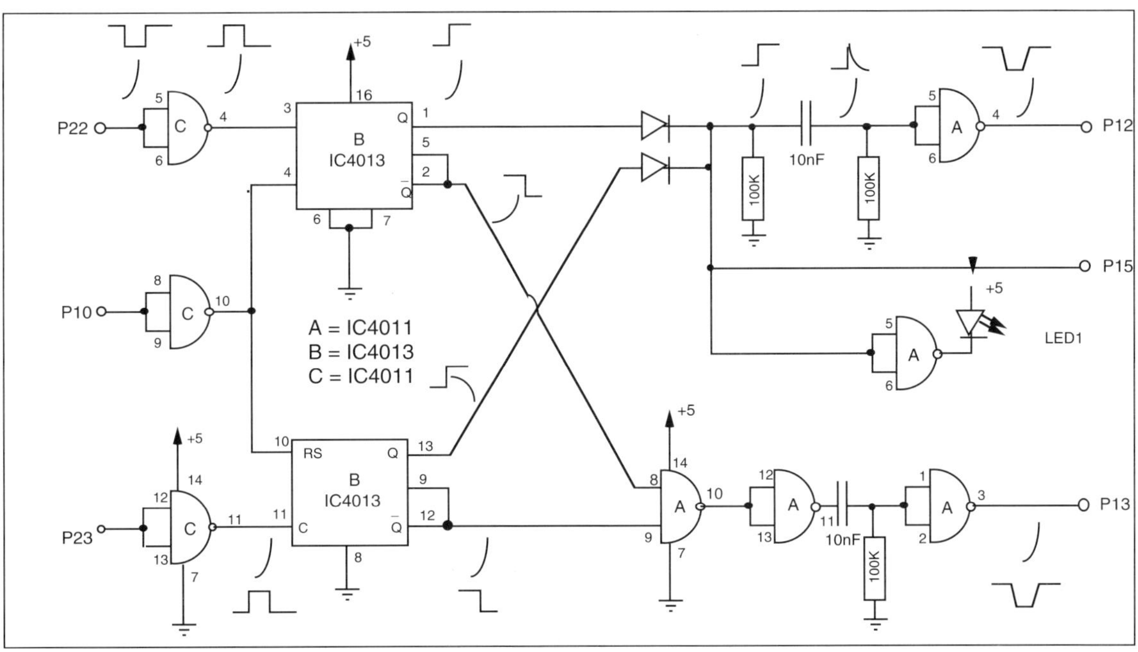

MODULE C — SCHEMATIC
The C module contains much blocking circuitry that prevents program collisions and missed initiating when several trains are in the detection block. Connection and LED function are described in a separate chart.

Double track overview

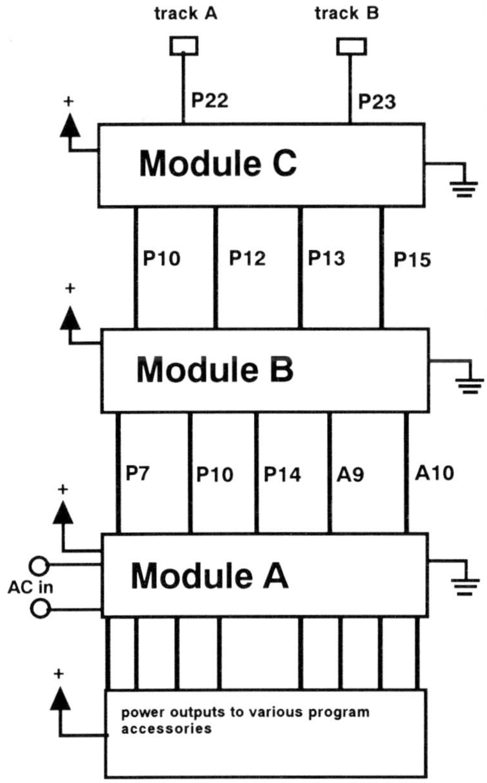

This is an overview of the double track version, module connections. All power feed leads can be connected to the base module A's voltage rectifier. Five volts is also available on module A at IC 7805, pin 3.

Connection Chart

A9:	address line in at EPROM 2716(EPROM) — module A
A10:	address line in at EPROM 2716(EPROM) — module A
LED1:	train in block & restart if new train enters block prior to termination of subprogram 3
LED2:	program indicator that lights when A10 is 1 (high level) – see program flow table
P10:	reset at startup
P12:	train entering block on single track with one-way traffic
P13:	train leaving block on single track with one-way traffic
P22:	train entering block on double track with bi-directional traffic – active as long as train in track A block
P23:	same as P22 but for track B block
P15:	subprogram 2 restart if new train enters block prior to termination of subprogram 3
P7:	program reset when subprogram 3 terminates
P14:	connection for reset and other units

Control of "soft" light

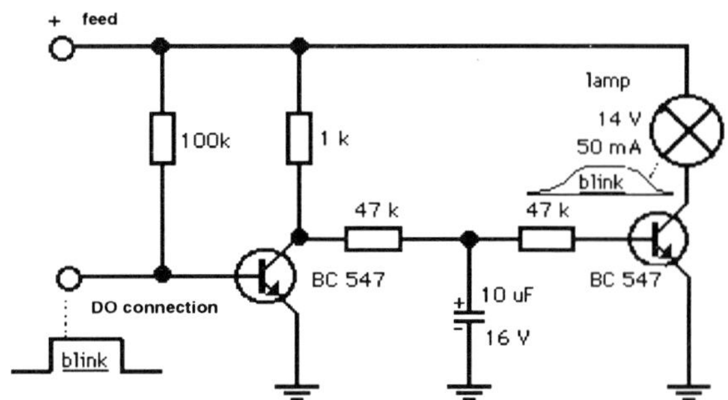

This circuits transforms abrupt on and off lamp switchings to gradual transitions from light to dark. For applications appropriate for soft light switching, the effect is dramatic.

Control of large bell

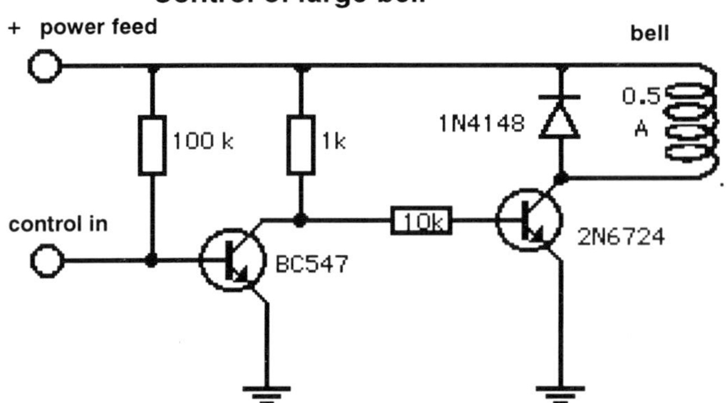

To operate a larger bell, the drive capacity of the smaller circuit is insufficient. This is a heavy duty variation that can easily handle 0.5A loads. The 2N6724 is a Darlington transistor but it can be replaced with the larger BD679.

Control of small bell

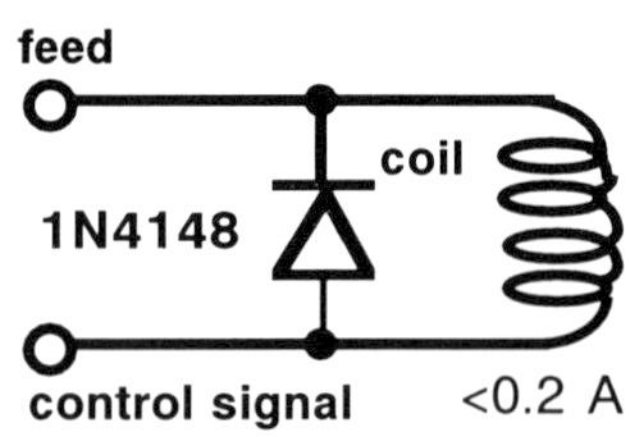

Brawa's small bell can be used with this circuit consisting of just one diode. The diode protects the driver circuits on the base module from transients and incorrect connections.

Fulgurex makes a switching relay for motor control. These are often used for slow motion track switch control. It is perfect for powering crossing gates.

Note that the motor regulation circuit consists of an absolute minimum of components. At first glance, it may look as if it wouldn't work at all. I recommend putting together a little test circuit before building the finished product just for demonstration purposes. Normally, a protective circuit is used at control connection points D5 and D6 so that the two circuit halves can never be activated simultaneously. I have omitted this protective circuit. A correct EPROM control program makes improper activation impossible. A dry run with two LEDs at outputs D5 and D6 will show that the outputs are never activated at the same time. A photo of my prototype circuit with the motor and gate mechanism is provided.

Lastly, a few words on program code... The program takes considerably longer to code than earlier programs. This is primarily because the automatic mechanism frequency must be higher than on earlier projects and that the activation time (meaning the time that the train is on the crossing trackage - between reed switches) is long. It's worth mentioning that the code should be entered into a 2716 EPROM in sequence, allowing program selection via control of addresses A9 and A10.

Connection of warning lights

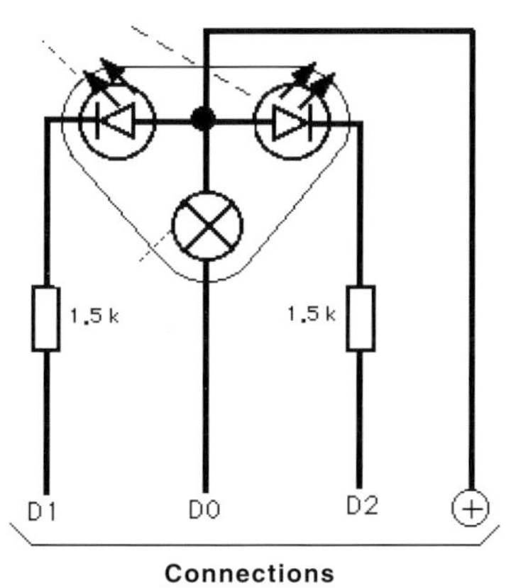

Regulating LEDs and lamps from a single output is easy using a common plus and series resistors.

Control of gate mechanism

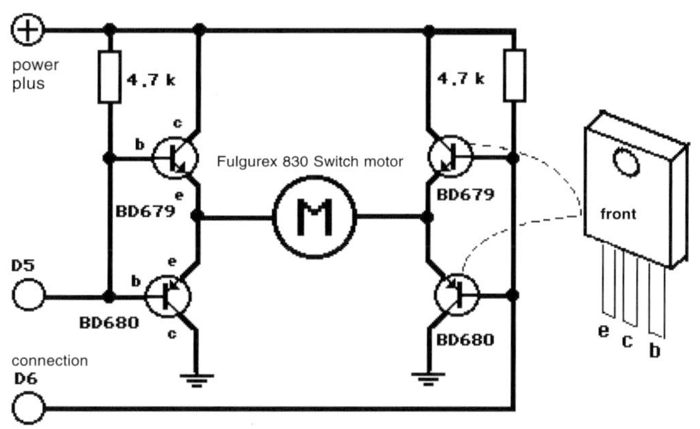

Motor control can't be any simpler than this. On and off, and reversing functions can be controlled using this circuit with four power transistors.

Eprom grade crossing – part 1

Source code in order of operation

```
81 81 80 80 81 81 80 80
81 81 80 80 81 81 80 80
81 81 80 80 81 81 80 80
81 81 80 80 81 81 80 80
81 81 80 80 81 81 80 80
81 81 80 80 81 81 80 80
81 81 80 80 81 81 80 80
81 81 80 80 81 81 80 80
81 81 80 80 81 81 80 80
81 81 80 80 81 81 80 80
81 81 80 80 81 81 80 80
81 81 80 80 81 81 80 80
81 81 80 80 81 81 80 80
81 81 80 80 81 81 80 80
81 81 80 80 81 81 80 80
81 81 80 80 81 81 80 80
81 81 80 80 81 81 80 80
81 81 80 80 81 81 80 80
81 81 80 80 81 81 80 80
81 81 80 80 81 81 80 80
81 81 80 80 81 81 80 80
81 81 80 80 81 81 80 80
81 81 80 80 81 81 80 80
81 81 80 80 81 81 80 80
81 81 80 80 81 81 80 80
81 81 80 80 81 81 80 80
81 81 80 80 81 81 80 80
81 81 80 80 81 81 80 80
81 81 80 80 81 81 80 80
81 81 80 80 81 81 80 80
81 81 80 80 81 81 80 80
81 81 80 80 81 81 80 80
81 81 80 80 81 81 80 80
81 81 80 80 81 81 80 80
81 81 80 80 81 81 80 80
81 81 80 80 81 81 80 80
81 81 80 80 81 81 80 80
81 81 80 80 81 81 80 80
81 81 80 80 81 81 80 80
81 81 80 80 81 81 80 80
81 81 80 80 81 81 80 80
81 81 80 80 81 81 80 80
81 81 80 80 81 81 80 80
81 81 80 80 81 81 80 80
81 81 80 80 81 81 80 80
81 81 80 80 81 81 80 80
81 81 80 80 81 81 80 80
81 81 80 80 81 81 80 80
81 81 80 80 81 81 80 80
81 81 80 80 81 81 80 80
81 81 80 80 81 81 80 80
81 81 80 80 81 81 80 80
81 81 80 80 81 81 80 80
81 81 80 80 81 81 80 80
81 81 80 80 81 81 80 80
81 81 80 80 81 81 80 80
81 81 80 80 81 81 80 80
81 81 80 80 81 81 80 80
81 81 80 80 81 81 80 80
81 81 80 80 81 81 80 80
81 81 80 80 81 81 80 80
81 81 80 80 81 81 80 80
81 81 80 80 81 81 80 80
81 81 80 80 81 81 80 80
```

```
8A 8C 92 94 8A 8C 92 94
8A 8C B2 B4 AA AC B2 B4
AA AC B2 B4 AA AC B2 B4
AA AC B2 B4 AA AC 82 84
82 84 82 84 82 84 82 84
82 84 82 84 82 84 82 84
82 84 82 84 82 84 82 84
82 84 82 84 82 84 82 84
82 84 82 84 82 84 82 84
82 84 82 84 82 84 82 84
82 84 82 84 82 84 82 84
82 84 82 84 82 84 82 84
82 84 82 84 82 84 82 84
82 84 82 84 82 84 82 84
82 84 82 84 82 84 82 84
82 84 82 84 82 84 82 84
82 84 82 84 82 84 82 84
82 84 82 84 82 84 82 84
82 84 82 84 82 84 82 84
82 84 82 84 82 84 82 84
82 84 82 84 82 84 82 84
82 84 82 84 82 84 82 84
82 84 82 84 82 84 82 84
82 84 82 84 82 84 82 84
82 84 82 84 82 84 82 84
82 84 82 84 82 84 82 84
82 84 82 84 82 84 82 84
82 84 82 84 82 84 82 84
82 84 82 84 82 84 82 84
82 84 82 84 82 84 82 84
82 84 82 84 82 84 82 84
82 84 82 84 82 84 82 84
82 84 82 84 82 84 82 84
82 84 82 84 82 84 82 84
82 84 82 84 82 84 82 84
82 84 82 84 82 84 82 84
82 84 82 84 82 84 82 84
82 84 82 84 82 84 82 84
82 84 82 84 82 84 82 84
82 84 82 84 82 84 82 84
82 84 82 84 82 84 82 84
82 84 82 84 82 84 82 84
82 84 82 84 82 84 82 84
82 84 82 84 82 84 82 84
82 84 82 84 82 84 82 84
82 84 82 84 82 84 82 84
82 84 82 84 82 84 82 84
82 84 82 84 82 84 82 84
82 84 82 84 82 84 82 84
82 84 82 84 82 84 82 84
82 84 82 84 82 84 82 84
82 84 82 84 82 84 82 84
82 84 82 84 82 84 82 84
82 84 82 84 82 84 82 84
82 84 82 84 82 84 82 84
82 84 82 84 82 84 82 84
82 84 82 84 82 84 82 84
82 84 82 84 82 84 82 84
82 84 82 84 82 84 82 84
82 84 82 84 82 84 82 84
82 84 82 84 82 84 82 84
82 84 82 84 82 84 82 84
82 84 82 84 82 84 82 84
82 84 82 84 82 84 82 84
```

Source code in order of operation

```
80 80 80 80 80 80 80 80
80 80 80 80 80 80 80 80
80 80 80 80 80 80 80 80
80 80 80 80 80 80 80 80
80 80 80 80 80 80 80 80
80 80 80 80 80 80 80 80
80 80 80 80 80 80 80 80
80 80 80 80 80 80 80 80
80 80 80 80 80 80 80 80
80 80 80 80 80 80 80 80
80 80 80 80 80 80 80 80
80 80 80 80 80 80 80 80
80 80 80 80 80 80 80 80
80 80 80 80 80 80 80 80
80 80 80 80 80 80 80 80
80 80 80 80 80 80 80 80
80 80 80 80 80 80 80 80
80 80 80 80 80 80 80 80
80 80 80 80 80 80 80 80
80 80 80 80 80 80 80 80
80 80 80 80 80 80 80 80
80 80 80 80 80 80 80 80
80 80 80 80 80 80 80 80
80 80 80 80 80 80 80 80
80 80 80 80 80 80 80 80
80 80 80 80 80 80 80 80
80 80 80 80 80 80 80 80
80 80 80 80 80 80 80 80
80 80 80 80 80 80 80 80
80 80 80 80 80 80 80 80
80 80 80 80 80 80 80 80
80 80 80 80 80 80 80 80
80 80 80 80 80 80 80 80
80 80 80 80 80 80 80 80
80 80 80 80 80 80 80 80
```

```
80 80 80 80 80 80 80 80
80 80 80 80 80 80 80 80
80 80 80 80 80 80 80 80
80 80 80 80 80 80 80 80
80 80 80 80 80 80 80 80
80 80 80 80 80 80 80 80
80 80 80 80 80 80 80 80
80 80 80 80 80 80 80 80
80 80 80 80 80 80 80 80
80 80 80 80 80 80 80 80
80 80 80 80 80 80 80 80
80 80 80 80 80 80 80 80
80 80 80 80 80 80 80 80
80 80 80 80 80 80 80 80
80 80 80 80 80 80 80 80
80 80 80 80 80 80 80 80
80 80 80 80 80 80 80 80
80 80 80 80 80 80 80 80
80 80 80 80 80 80 80 80
80 80 80 80 80 80 80 80
80 80 80 80 80 80 80 80
80 80 80 80 80 80 80 80
80 80 80 80 80 80 80 80
80 80 80 80 80 80 80 80
80 80 80 80 80 80 80 80
80 80 80 80 80 80 80 80
80 80 80 80 80 80 80 80
80 80 80 80 80 80 80 80
80 80 80 80 80 80 80 80
82 84 C2 C4 C2 C4 C2 C4
C2 C4 C2 C4 C2 C4 C2 C4
C2 C4 C2 C4 C2 C4 82 84
82 84 82 84 82 04 02 04
02 04 02 04 02 04 02 04
00 00 00 00 00 00 00 00
00 00 00 00 00 00 00 00
```

MINI-TIP

End of train warning device

This little project arose from needs in HO scale. Afterwards, I realized that the five components and LEDs that make up the circuit could be assembled into a unit so compact that it could be used in N-scale. In HO scale, use 3mm LEDs and in N scale, 1mm LEDs.

We'll build the unit so that it can be used for both AC and DC operations - that's why the rectifier is included. There is also a voltage stabilizer IC that delivers a constant 5 volts regardless of considerably higher track voltages. IC 3909 is a flasher that will operate on as little as 1.5V when connected in the normal manner. The more experienced among us can get by on even less by taking advantage of the voltage drop over 2-3 diodes (for example, 1N4148) wired in series. This provides 1.4-2.1 volts. Power feed is at pin 5 for plus and pin 4 for ground.

The LEDs (I've modeled the two-light devices used in Sweden) are parallel connected between pins 6 and 8 and inserted in holes drilled at the rear of the car. The values of the electrolytic capacitors are not critical but the flash frequency slows with capacitance higher than that shown between pins 2 and 8. Note that if IC 3909 is connected directly to a power source (such as a battery), voltage should not exceed 6VDC.

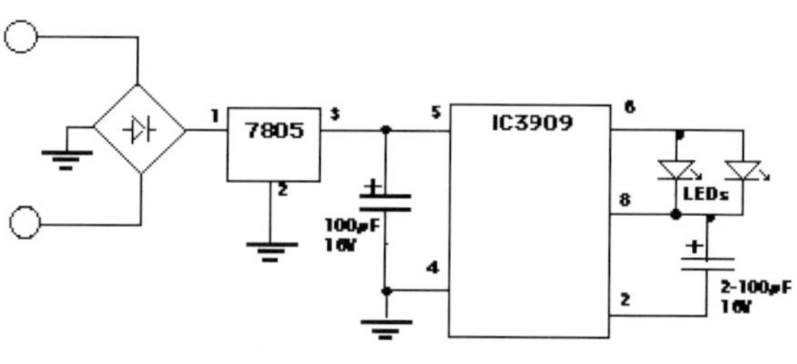

More projects using the program module

The smart grade crossing adapted for Märklin H0

It is a good idea to complement the general description of grade crossings with a section devoted to a Märklin application. This is partially because Märklin is popular and an important advocate of model railroading but also because the brand has its own profile - AC operation the most noteworthy characteristic. Furthermore, Märklin offers a wide selection of mechanized grade crossings, remote controlled by electromagnets.

Märklin modelers have two types of track to choose among - M- and K-track. These are two variations on the same theme. Grade crossings are specific for each of these. I'll be centering on the classic, stamped metal M-track versions . Sorry, K-track modelers - you will have to substitute type numbers when reading the description.

There is a 100% mechanically regulated crossing (7390 M) that is activated by train weight. We will eliminate this crossing from further discussion and relegate it to the toy box.

BACKWARDS AND WITH A SINGLE OUTPUT

If you've read the previous, general description of grade crossings, you'll remember that the gate control motor spun in one direction to lower the gates and in the other direction to raise them. Control signals were via D5 and D6.

The Märklin gates are held down by continuos activation of an electromagnet. When current is interrupted, the gates slam to the up position. This is the opposite of the previous version - "off" now means raised gates while "on" means lowered gates.

If a semi-direct control method is acceptable, a simpler electric servo can be connected to D6. When D6 is activated,

Special A

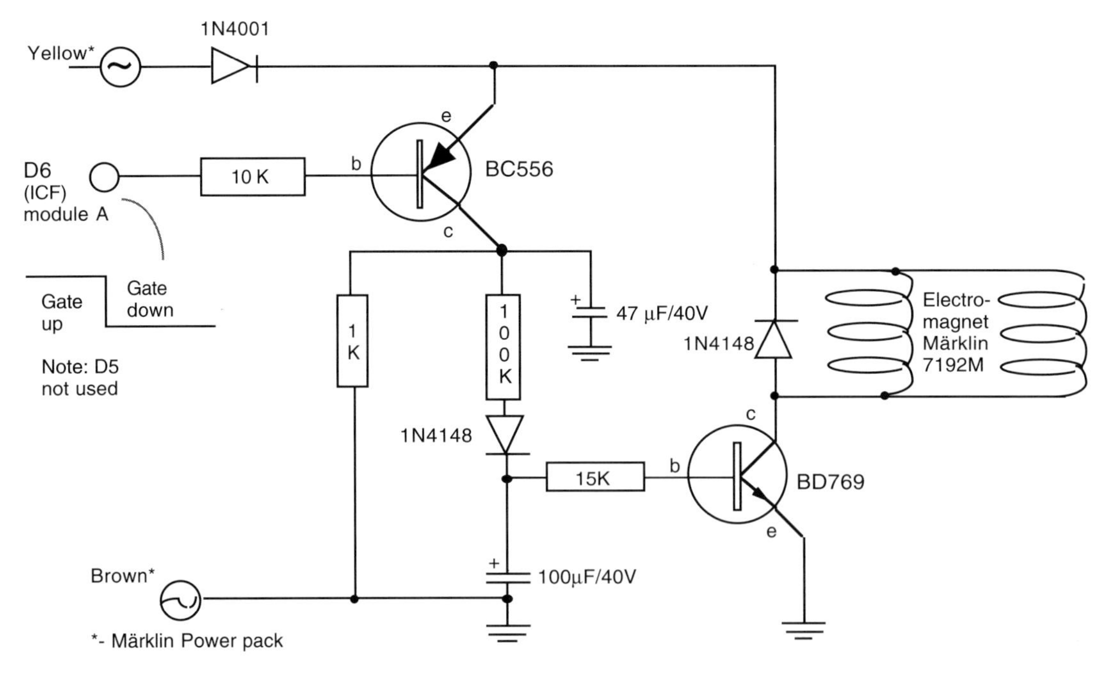

The simple way of connecting this module is to hook it up to a Märklin transformer according to the adjacent information. In order to be able to use a transistor control for this module there is a rectifying diode 1N4001 from the "yellow" connector.

As an alternative you may also use a Triac (a double thyristor) which opens up the possibility to keep the AC power. You should be aware of the electromagnetic design of this crossing, reacting directly and closing the gate when a control pulse on the D6 pin is generated.

This Märklin railroad crossing device is designed in another way than common crossings. Therefor it needs some special code instead of that generated in the last chapter. One change is that only one output is used (D6) to close the gates and when this output goes low again the gates will be opened.

Another change is the quick reaction. This code needs to take in consideration the extremely fast reaction and therefor start opening the gates much later than usually. The sub program code in this special control program has though still the same start address.

the gates are lowered and remain so as long as activation current is present.

Direct control is most suitable for crossing number 7292 M. It features double warning signals on each side of the tracks. It can also be complemented for double track as mentioned in the previous section. And of course we'll be using the basic modules A, B and C for the Märklin crossing as well.

We must adapt the servo units and program code to these special crossings. If the warning lights are to blink alternately, they must be freed from their common connection point with the magnet winding. Instead, connect the lamps to D1 and D2 and the magnet winding to D6. The common return lead is connected to plus.

A "program zero" in the control code for D6 lowers the gates. This is because IC F does not, in this state, ground the effect transistor in our servo circuit. Instead, the transistor

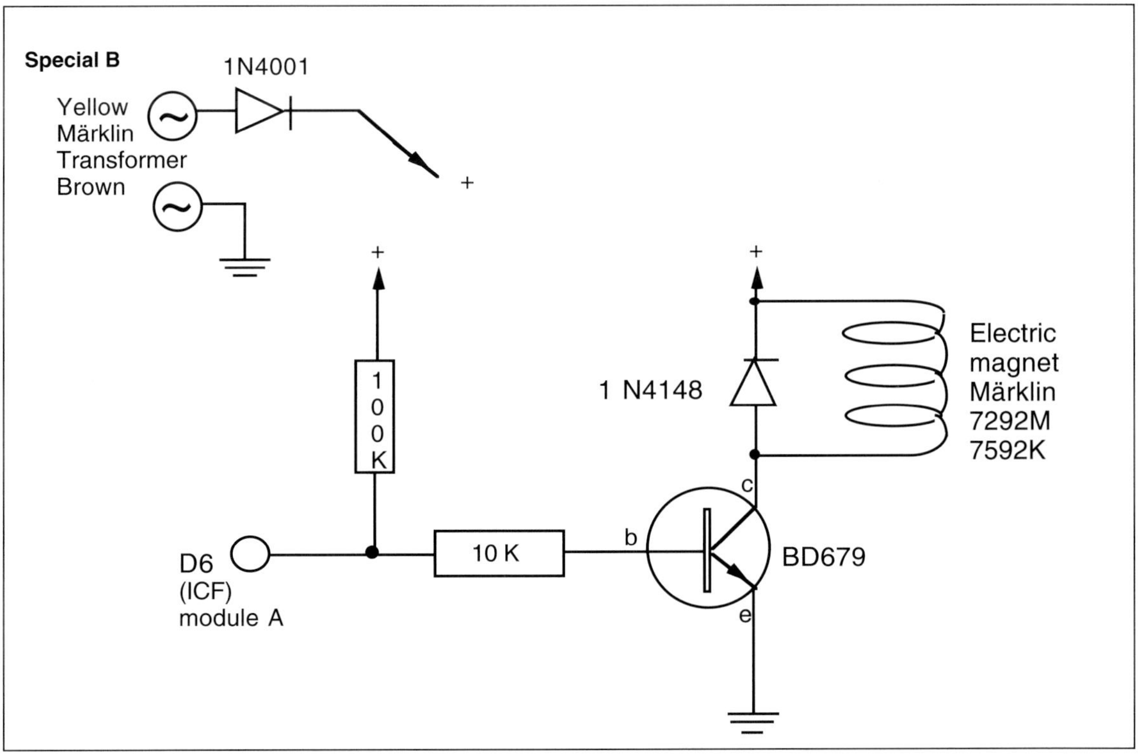

This is how you connect the Märklin 7192 crossing module. Also here the basic motion is generated via an electric magneto, but there is a gear driveline in between, helping the gates to move more soft than earlier examples.

Through several experiments this is the final proposal to generate as soft as possible motion (which in fact is a lot better than original operation). If you have an original Märklin transformer there is one more way to make fine adjustment. Change the yellow feeder to red on the rear side of the transformer and adjust the output voltage via the speedknob. This will give you the best individual adjustment of your gates.

row 1 - 32

81	81	80	80	81	81	80	80
81	81	80	80	81	81	80	80
81	81	80	80	81	81	80	80
81	81	80	80	81	81	80	80
81	81	80	80	81	81	80	80
81	81	80	80	81	81	80	80
81	81	80	80	81	81	80	80
81	81	80	80	81	81	80	80
81	81	80	80	81	81	80	80
81	81	80	80	81	81	80	80
81	81	80	80	81	81	80	80
81	81	80	80	81	81	80	80
81	81	80	80	81	81	80	80
81	81	80	80	81	81	80	80
81	81	80	80	81	81	80	80
81	81	80	80	81	81	80	80
81	81	80	80	81	81	80	80
81	81	80	80	81	81	80	80
81	81	80	80	81	81	80	80
81	81	80	80	81	81	80	80
81	81	80	80	81	81	80	80
81	81	80	80	81	81	80	80
81	81	80	80	81	81	80	80
81	81	80	80	81	81	80	80
81	81	80	80	81	81	80	80
81	81	80	80	81	81	80	80
81	81	80	80	81	81	80	80
81	81	80	80	81	81	80	80
81	81	80	80	81	81	80	80
81	81	80	80	81	81	80	80
81	81	80	80	81	81	80	80
81	81	80	80	81	81	80	80

row 33 - 64

81	81	80	80	81	81	80	80
81	81	80	80	81	81	80	80
81	81	80	80	81	81	80	80
81	81	80	80	81	81	80	80
81	81	80	80	81	81	80	80
81	81	80	80	81	81	80	80
81	81	80	80	81	81	80	80
81	81	80	80	81	81	80	80
81	81	80	80	81	81	80	80
81	81	80	80	81	81	80	80
81	81	80	80	81	81	80	80
81	81	80	80	81	81	80	80
81	81	80	80	81	81	80	80
81	81	80	80	81	81	80	80
81	81	80	80	81	81	80	80
81	81	80	80	81	81	80	80
81	81	80	80	81	81	80	80
81	81	80	80	81	81	80	80
81	81	80	80	81	81	80	80
81	81	80	80	81	81	80	80
81	81	80	80	81	81	80	80
81	81	80	80	81	81	80	80
81	81	80	80	81	81	80	80
81	81	80	80	81	81	80	80
81	81	80	80	81	81	80	80
81	81	80	80	81	81	80	80
81	81	80	80	81	81	80	80
81	81	80	80	81	81	80	80
81	81	80	80	81	81	80	80
81	81	80	80	81	81	80	80
81	81	80	80	81	81	80	80
81	81	80	80	81	81	80	80

base retains plus potential until a coded "one" activates IC F, which in turn, grounds the effect transistor base. Although this is handled by the furnished code (read how it is managed in the previous section covering program modules), it can be good to know if you want to make adjustments in the code or write your own programs.

SLOW MOTION

There is a slightly more expensive crossing numbered 7192 M. Unfortunately, it lacks flashing lights (you can always add your own) but has a better gate mechanism. It uses a cogged member from the electromagnet that raises and lowers the gates both slower and smoother. In this case, it can be worthwhile to invest in a little more control electronics (especially module A) to emphasize the slow, gate movement. The crossing is shown in the photo along with the specially programmed EPROM. With the addition of an optional accessory, the crossing can also be used with double track.

The reason for writing a program completely different from the previous one has actually been mentioned earlier - continual activation of D6 while trains are passing demands changes in code. Times for raising and lowering the gates have also been altered somewhat.

If you are wondering how far from the crossing the reed switches should be mounted, I can only say that generally, it depends on how fast your trains run and how you have adjusted the trim potentiometer on the base module. On my test layout, the distance was just under seven feet in both directions.

row 65 - 96

8A	8C	92	94	8A	8C	92	94
8A	8C	D2	D4	CA	CC	D2	D4
CA	CC	D2	D4	CA	CC	D2	D4
C2	C4	C2	C4	C2	C4	C2	C4
C2	C4	C2	C4	C2	C4	C2	C4
C2	C4	C2	C4	C2	C4	C2	C4
C2	C4	C2	C4	C2	C4	C2	C4
C2	C4	C2	C4	C2	C4	C2	C4
C2	C4	C2	C4	C2	C4	C2	C4
C2	C4	C2	C4	C2	C4	C2	C4
C2	C4	C2	C4	C2	C4	C2	C4
C2	C4	C2	C4	C2	C4	C2	C4
C2	C4	C2	C4	C2	C4	C2	C4
C2	C4	C2	C4	C2	C4	C2	C4
C2	C4	C2	C4	C2	C4	C2	C4
C2	C4	C2	C4	C2	C4	C2	C4
C2	C4	C2	C4	C2	C4	C2	C4
C2	C4	C2	C4	C2	C4	C2	C4
C2	C4	C2	C4	C2	C4	C2	C4
C2	C4	C2	C4	C2	C4	C2	C4
C2	C4	C2	C4	C2	C4	C2	C4
C2	C4	C2	C4	C2	C4	C2	C4
C2	C4	C2	C4	C2	C4	C2	C4
C2	C4	C2	C4	C2	C4	C2	C4
C2	C4	C2	C4	C2	C4	C2	C4
C2	C4	C2	C4	C2	C4	C2	C4
C2	C4	C2	C4	C2	C4	C2	C4
C2	C4	C2	C4	C2	C4	C2	C4
C2	C4	C2	C4	C2	C4	C2	C4
C2	C4	C2	C4	C2	C4	C2	C4
C2	C4	C2	C4	C2	C4	C2	C4
C2	C4	C2	C4	C2	C4	C2	C4

row 97- 128

C2	C4	C2	C4	C2	C4	C2	C4
C2	C4	C2	C4	C2	C4	C2	C4
C2	C4	C2	C4	C2	C4	C2	C4
C2	C4	C2	C4	C2	C4	C2	C4
C2	C4	C2	C4	C2	C4	C2	C4
C2	C4	C2	C4	C2	C4	C2	C4
C2	C4	C2	C4	C2	C4	C2	C4
C2	C4	C2	C4	C2	C4	C2	C4
C2	C4	C2	C4	C2	C4	C2	C4
C2	C4	C2	C4	C2	C4	C2	C4
C2	C4	C2	C4	C2	C4	C2	C4
C2	C4	C2	C4	C2	C4	C2	C4
C2	C4	C2	C4	C2	C4	C2	C4
C2	C4	C2	C4	C2	C4	C2	C4
C2	C4	C2	C4	C2	C4	C2	C4
C2	C4	C2	C4	C2	C4	C2	C4
C2	C4	C2	C4	C2	C4	C2	C4
C2	C4	C2	C4	C2	C4	C2	C4
C2	C4	C2	C4	C2	C4	C2	C4
C2	C4	C2	C4	C2	C4	C2	C4
C2	C4	C2	C4	C2	C4	C2	C4
C2	C4	C2	C4	C2	C4	C2	C4
C2	C4	C2	C4	C2	C4	C2	C4
C2	C4	C2	C4	C2	C4	C2	C4
C2	C4	C2	C4	C2	C4	C2	C4
C2	C4	C2	C4	C2	C4	C2	C4
C2	C4	C2	C4	C2	C4	C2	C4
C2	C4	C2	C4	C2	C4	C2	C4
C2	C4	C2	C4	C2	C4	C2	C4
C2	C4	C2	C4	C2	C4	C2	C4
C2	C4	C2	C4	C2	C4	C2	C4
C2	C4	C2	C4	C2	C4	C2	C4

A common baseline for all the road x-ing projects is durability. In order to get certain sequences to be carried out it is necessary that the trains are well maintained and run under clean conditions. A train that stops unintentionally because of bad current pick-up in the active x-ing section is electrically speaking a "non train" since there is no track detection finding it.

You can avoid that by keeping dirt away and by keeping the critical electrical components of the rolling vehicles in a "mint condition". This is one example. A new pick-up shoe for a Märklin engine is a priceworthy investment instead of trying to grind the old and worn out one.

Source code in order of operation

3.

```
80 80 80 80 80 80 80 80    80 80 80 80 80 80 80 80
80 80 80 80 80 80 80 80    80 80 80 80 80 80 80 80
80 80 80 80 80 80 80 80    80 80 80 80 80 80 80 80
80 80 80 80 80 80 80 80    80 80 80 80 80 80 80 80
80 80 80 80 80 80 80 80    80 80 80 80 80 80 80 80
80 80 80 80 80 80 80 80    80 80 80 80 80 80 80 80
80 80 80 80 80 80 80 80    80 80 80 80 00 80 80 80
80 80 80 80 80 80 80 80    00 00 80 80 00 00 80 00
80 80 80 80 80 80 80 80    80 80 80 80 80 80 80 80
80 80 80 80 80 80 80 80    80 80 80 80 80 80 80 80
80 80 80 80 80 80 80 80    80 80 80 80 80 80 80 80
80 80 80 80 80 80 80 80    80 80 80 80 80 80 80 80
80 80 80 80 80 80 80 80    80 80 80 80 80 80 80 80
80 80 80 80 80 80 80 80    80 80 80 80 80 80 80 80
80 80 80 80 80 80 80 80    80 80 80 80 80 80 80 80
80 80 80 80 80 80 80 80    80 80 80 80 80 80 80 80
80 80 80 80 80 80 80 80    80 80 80 80 80 80 80 80
80 80 80 80 80 80 80 80    80 80 80 80 80 80 80 80
80 80 80 80 80 80 80 80    80 80 80 80 80 80 80 80
```

4.

```
80 80 80 80 80 80 80 80    80 80 80 80 80 80 80 80
80 80 80 80 80 80 80 80    80 80 80 80 80 80 80 80
80 80 80 80 80 80 80 80    80 80 80 80 80 80 80 80
80 80 80 80 80 80 80 80    80 80 80 80 80 80 80 80
80 80 80 80 80 80 80 80    80 80 80 80 80 80 80 80
80 80 80 80 80 80 80 80    80 80 80 80 80 80 80 80
80 80 80 80 80 80 80 80    80 80 80 80 80 80 00 80
80 80 80 80 80 80 80 80    80 80 80 80 80 80 80 80
80 80 80 80 80 80 80 80    80 80 80 80 80 80 80 80
80 80 80 80 80 80 80 80    C2 C4 C2 C4 C2 C4 C2 C4
80 80 80 80 80 80 80 80    C2 C4 C2 C4 C2 C4 C2 C4
80 80 80 80 80 80 80 80    C2 C4 C2 C4 C2 C4 82 84
80 80 80 80 80 80 80 80    82 84 82 84 82 04 02 04
80 80 80 80 80 80 80 80    02 04 02 04 02 04 02 04
80 80 80 80 80 80 80 80    00 00 00 00 00 00 00 00
80 80 80 80 80 80 80 80    00 00 00 00 00 00 00 00
80 80 80 80 80 80 80 80
```

MINI-TIP

**Mini-tips are small, independent construction projects and light theoretical descriptions.
Mixing several small mini-tips with the heavier subjects in the first book met with favorable response.
Many readers have called or written, asking for more mini-tips.
Here and on the following pages is a new collection of mini-tips. Hopefully, you'll find just the thing for your own pike. Other mini-tips are scattered throughout the book.**

MINI-TIP

Speed regulation device for small stationary motors

On model railroads, you often find several different kinds of stationary, electric motors that need speed regulation. Typical examples are funicular railways and water wheels. There are even other complex motor circuits that can benefit from speed regulation - motor circuits that operate grade crossing gates and track switches are other examples.

Small electric motors (please note that this regulation circuit is for DC motors), requiring up to 0.5A can be controlled by a BD 679 transistor, for example. If the motor is to be run continuously (more or less), be sure to equip the transistor with a cooling flange. A good overheating indicator in cases like this is a finger, lightly pressed against the transistor casing. If the transistor is warm enough to burn your finger, more cooling is needed. Of course you could calculate the transistor's cooling requirements but this is not within the scope of this book.

In this example, the thermal load on the transistor is not especially large. This is because we regulate speed with pulses from IC 555 - high motor current is present only intermittently.

If the motor is of a larger type, use a TIP 41 or similar effect transistor. With a cooling flange, the circuit is sufficient for 3 amp operation.

Speed regulator for small motors

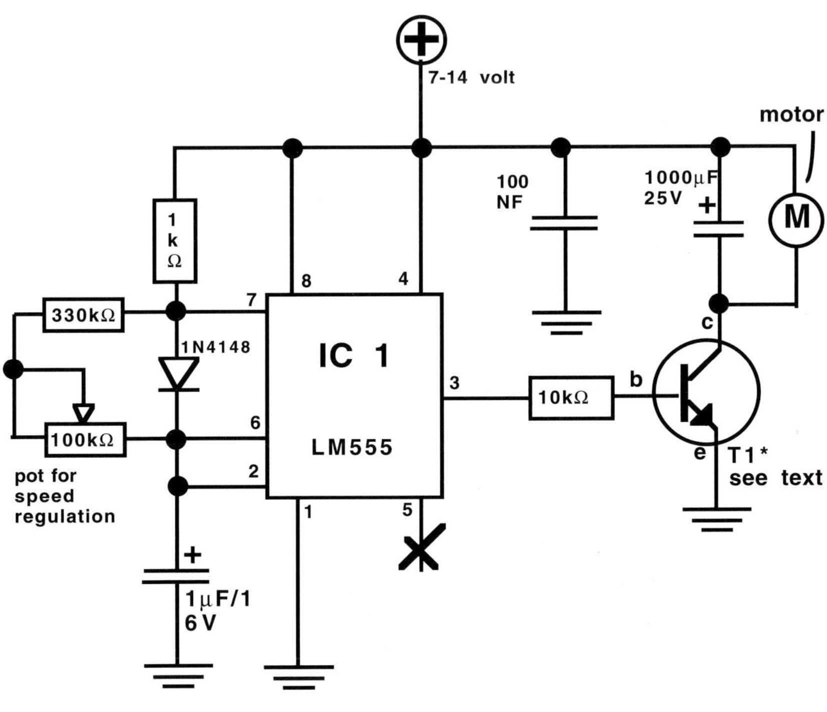

MINI-TIP

Power transistor protection

Model railroader use power transistors primarily in circuits such as controllers and electronic relays. In both cases, large unwanted currents, far in excess of those expected, can arise with short circuits and other overloads.

Here are two ways to increase the margin of safety for power transistors. In the example, a MJE3055 was chosen. This is because it is a large NPN transistor common to effect regulation circuits. In the circuit using three of transistors, all should be equipped with cooling flanges if current is large enough to warrant cooling in the basic construction. The other circuit has built-in current regulation so that MJE3055's maximum current limit is never reached.

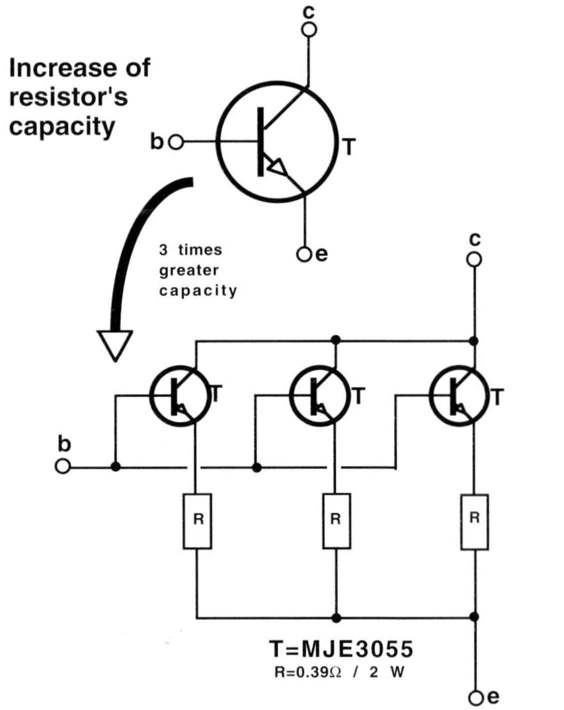

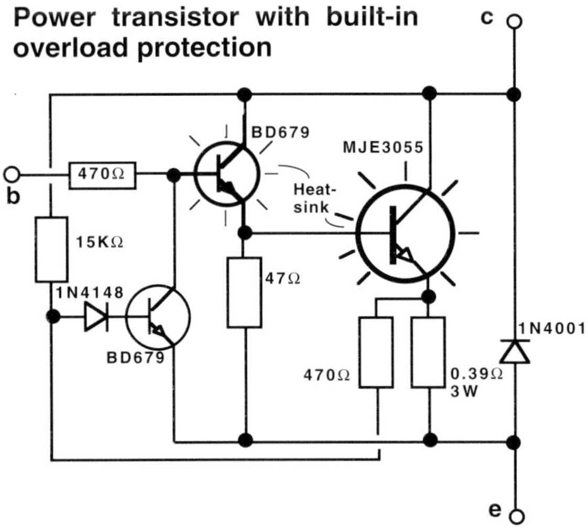

MINI-TIP:

Warning lights for roadwork

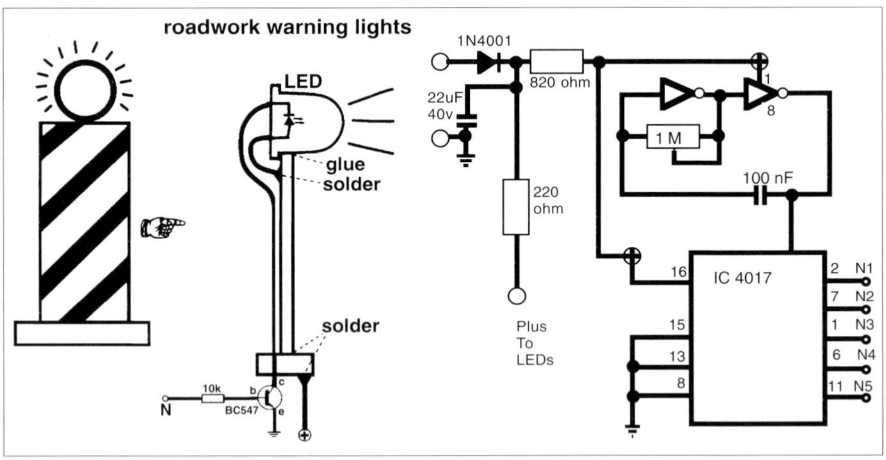

It's often worthwhile to create a small roadwork scene on the layout. There is an abundance of appropriate vehicles available — trucks, backhoes, etc.

Around the work place, warning signals should be placed to warn the "motorists". This little circuit does just that. One of the illustrations shows how to solder and glue together small brass posts for the warning lights. These can later be painted with orange and yellow stripes.

Depending upon modeling scale, use LEDs in sizes from 1mm to 5mm. The electronics consist of a three-component oscillator, a counter IC and transistor circuitry to each LED. The unit is either connected directly to a small battery or via a diode/resistor circuit to a power pack. The 220Ω resistor is intended for five LEDs. If two lights are use, change this to 680Ω. Adjust the frequency for the chase lights with a trim potentiometer.

MINI-TIP

Sound amplifier

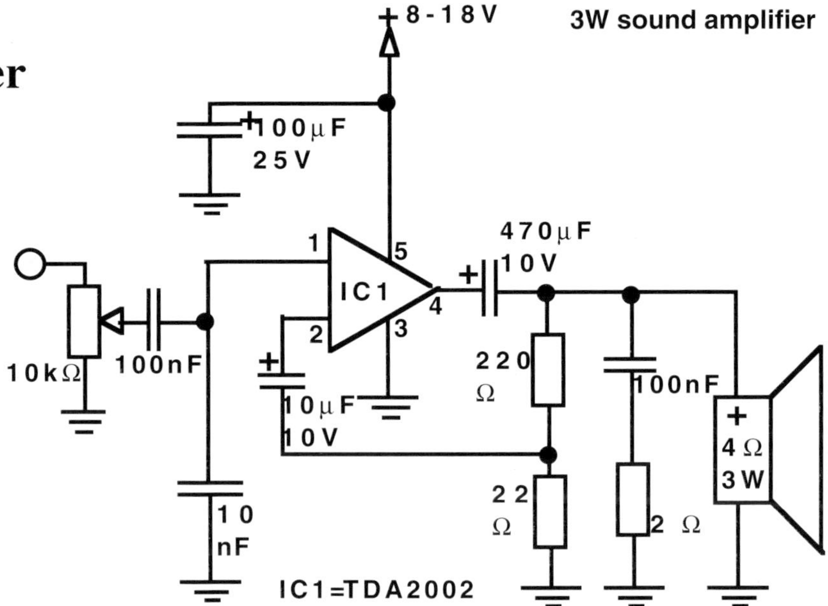

This amplifier is universal in that it can be used for many different sound sources and is compact enough that it can easily be placed onboard in both N and HO scale. Some sound effects can be located in buildings, making the amplifier usable for stationary, Z scale applications.

There are sound projects in the book that refer directly to this amplifier. The schematic shows an IC TDA 2002. It is an inexpensive and reliable, monolithic amplification circuit that can regulate a maximum effect of 8W. There is a newer version (completely pin compatible) in the same low price range - TDA 2003 with 10W capabilities, that we can also use.

Small, powerful speakers are available at good prices. Capable of handling 100mW and measuring about one inch in diameter, they are ideal for onboard sound effects. An N scale boxcar can easily swallow the entire sound unit.

The amplifier is assembled very compactly (omit the circuit board) with the help of tantalum electrolytes and with a 10kΩ volume potentiometer in sub-miniature format. Because the IC often has a much greater capacity than that recommended for the speaker, be careful when you power-up for the first time. Turn the potentiometer to minimum (measure resistance if you are uncertain) and slowly increase the volume after startup to the appropriate level. The IC itself is "idiot proof". It has built-in thermal protection, overvoltage protection and is short circuit proof.

MINI-TIP:

Warning cones with flashing lights

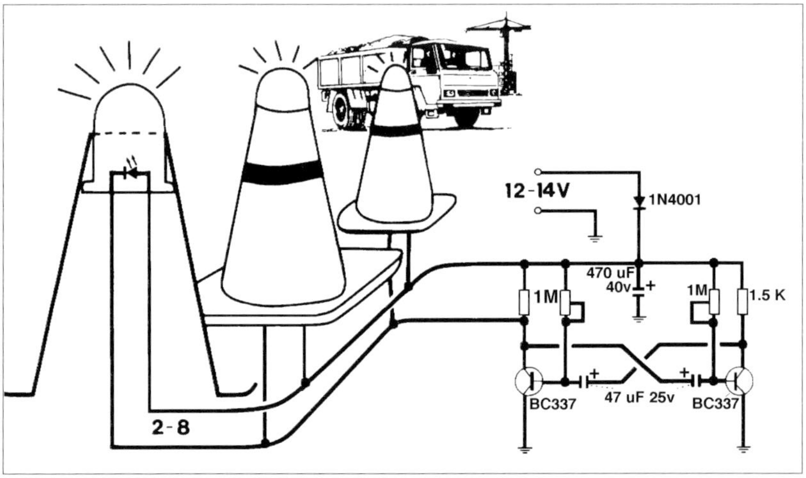

Orange warning cones with yellow reflective tape are common roadway features. They can be set up at traffic accidents or at roadwork sites when car traffic must be redirected. At night, flashing yellow lights are clamped to their tops for increased visibility.

These can be modeled rather easily on a train layout. Cones are available in both HO and N scales but can be built using mechanical pencil tips, minus the lead. An LED in a suitable size is glued in the top and leads run through the hole and the layout surface. The cones can be painted afterwards using modeling paint.

The circuit presented is meant for 2-8 cones, flashing with even intervals. Frequency is adjusted with a trim potentiometer. With a switch on the control panel, the module is connected to either a power pack or appropriate battery pack.

MINI-TIP

When I need an extra gate

Just as often one as has a spare gate or two in an IC – that case is covered in another mini-tip — the situation arises that an additional gate is an absolute necessity. I'm reluctant to use a whole IC for just one of its four gates. Here is alternative for two of the gate types. With a few diodes and a resistor, an extra gate can be assembled, saving space in a crowded area.

AND-gate substitute

A	B	C
0	0	0
0	1	0
1	0	0
1	1	1

AND-gate logic table

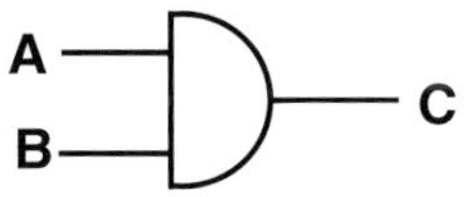

AND-gate symbol

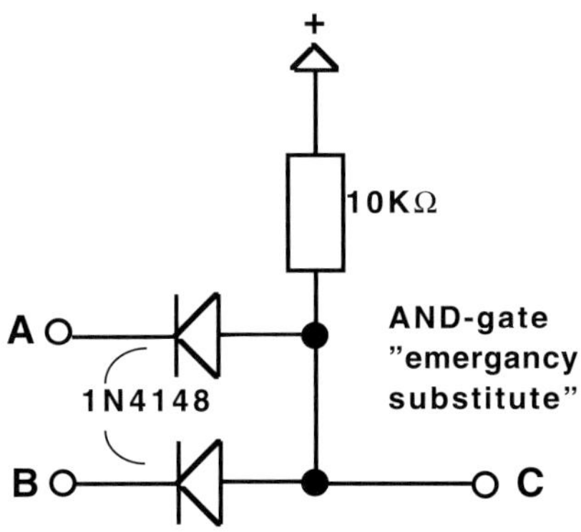

OR-gate substitute

A	B	C
0	0	0
0	1	1
1	0	1
1	1	1

OR-gate logic table

OR-gate symbol

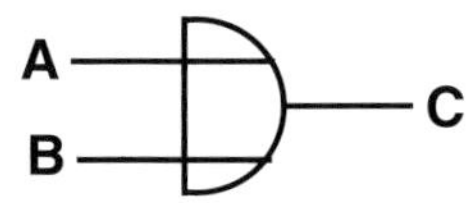

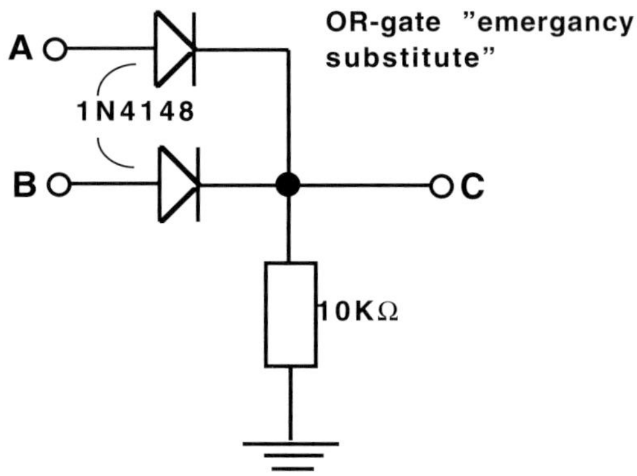

MINI-TIP

Double blink with six components

You can hardly get by simpler or cheaper. This little blink module is built around a standard IC with just two resistors and a capacitor. Either one or two LEDs are used as a blink unit. If two are used and they are connected as shown in the figure, they will flash alternately. Note that no series resistance for the LEDs is included. Their current is limited by the IC.

At 9V, the module draws approximately 10mA. Two of the four NAND gates in the IC form an oscillator and the other two drive circuits for the LEDs.

The stated component values provide a blink frequency of one per second. If the capacitor is changed to 10nF the blink frequency is doubled.

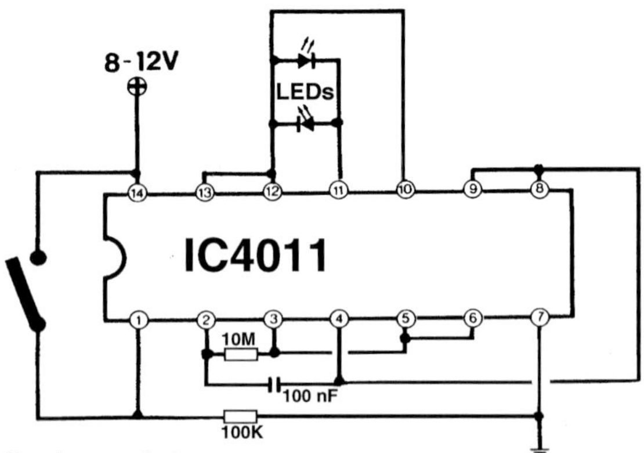

flash module

This module can easily be combined with other electronics. In that case, control circuitry replaces the switch at pin 1. A "one" at the pin starts the blinking.

MINI-TIP

An AC switch servo

The majority of the book's construction projects have featured direct current. For that reason, it can be appropriate to include an alternating current project.

The power pack accessory connection terminals are AC, whether you are running Märklin (a 100% AC system, with the exception of digital applications) or the more common DC systems. Electronic building blocks, such as switch controllers, are usually connected to the AC accessory terminals via a rectifier and filter (the filter uses a capacitor to even out DC voltage peaks). Effect transistors are then used to control the accessories.

In this project, we will use a thyristor to control switch coils with AC straight from the accessory terminals. Note that two-way switches normally have two coils (just like a bistable relay). This means that the setup shown in the figure will have to be duplicated for each coil. The circuit as shown in the figure could be used to control an electromagnetic uncoupler.

The NAND gate (one quarter of an IC 4011) in the figure functions in this circuit as an inverter. A momentary low pulse at its input results in a corresponding high pulse at its output to the thyristor. The IC can work within a voltage range of 3-15 volts regardless of which voltage in the 14-25 volt range is used on the AC side.

The thyristor's control input G (gate) triggers the ON function. Current begins to flow between the anode A and the cathode C, and through the coil. This activates the electromagnet. The thyristor leads current in the forward direction only. For this reason, it has a high inverse voltage (double working voltage) during the negative AC voltage period. The thyristor is switched off when this negative voltage appears but is switched on again during each positive period of the AC sine wave (~). This continues for as long as the high pulse at the NAND gate's output remains.

Suitable thyristors for currents of up to 0.5A are TIC45 or EC103. For larger currents, a C122 is recommended (maximum 8W and an inverse voltage of 200V). An important practical detail, noted in the figure, is that the AC "minus" must not be connected to the signal ground. There are voltage spikes in the AC portion of the circuit that have no place in a small-signal system.

Thyristor connection

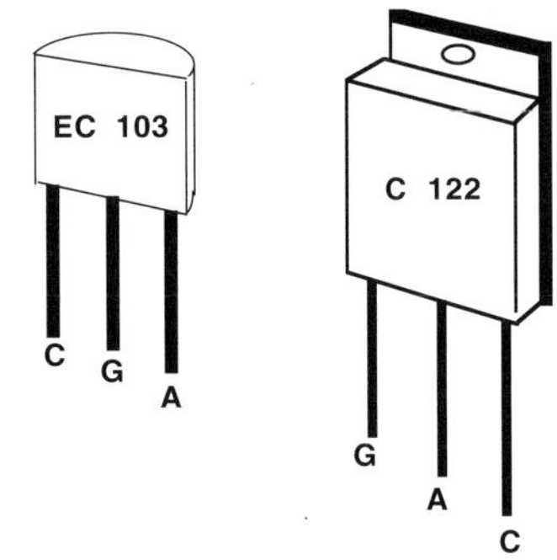

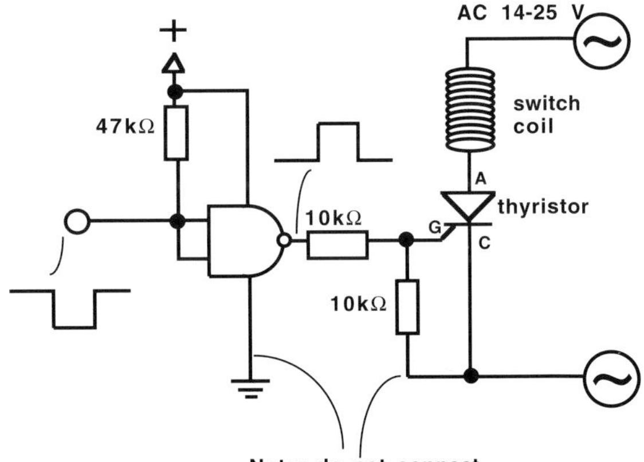

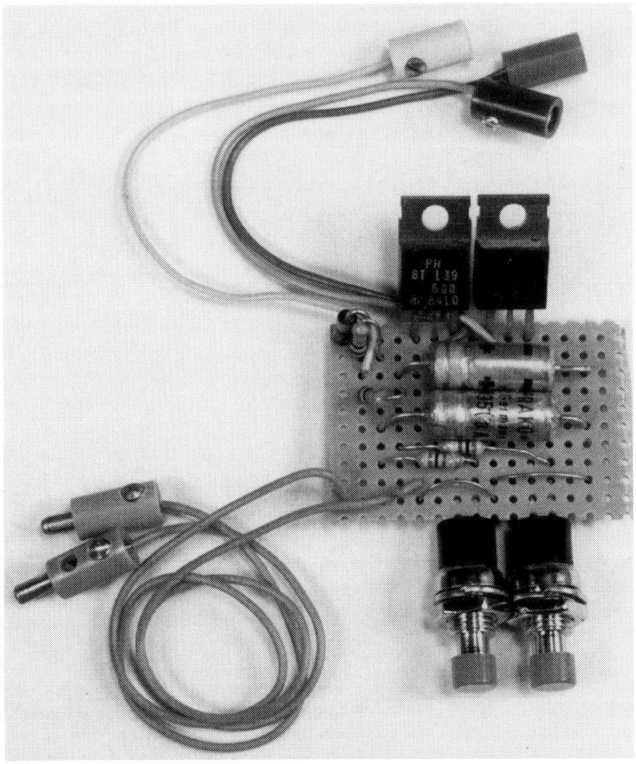

A simple and priceworthy turnout switch booster can either be based on tyristors or transitors. The elschematics for (half) a switch is shown here and most often it is practical to make twin units for every switch. This is a command control unit for dual electromagnetic switch, each half being operated by a buttonswitch. No cooling is normally needed unless you toggle a substancial amount of switches to be operated in parallel by the same power unit.

MINI-TIP

Turn your reverser right

Changing train direction on an analogue, DC layout is done by reversing polarity at the rails — the plus rail becomes minus and vice versa. Although this can done remotely using a relay, the usual method is with a DPDT switch on the power pack or control panel.

We'll limit ourselves to our own constructions, taking for granted that power pack manufactures have done things right.

Study the two figures, noting that the reversing switches are turned 180° in relation to each other. The figure without track shows the split second of the reversing maneuver when one of the spring loaded poles has already thrown and the second is soon to follow. Even though both rails are momentarily connected to one another, no damage is done in that there is no flow of current. If the reversing switch is connected as on the crossed-over figure, a brief short will occur. I should add that all approved power packs have overload protection and that this "little" short is marginal compared to derailments and tools forgotten on trackage. But one might as well do things right from the start.

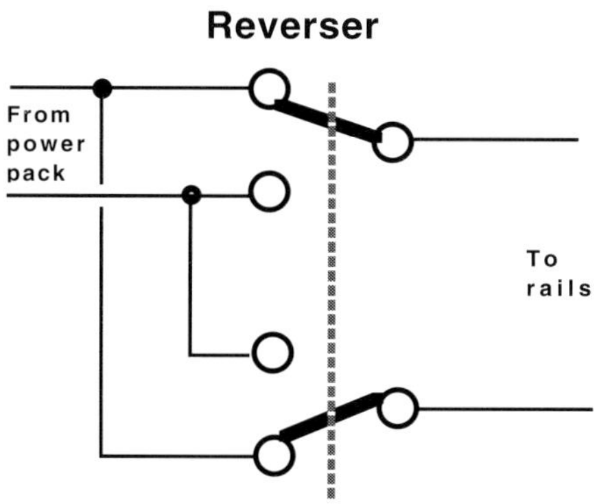

Contacts can take this position during reversing operation

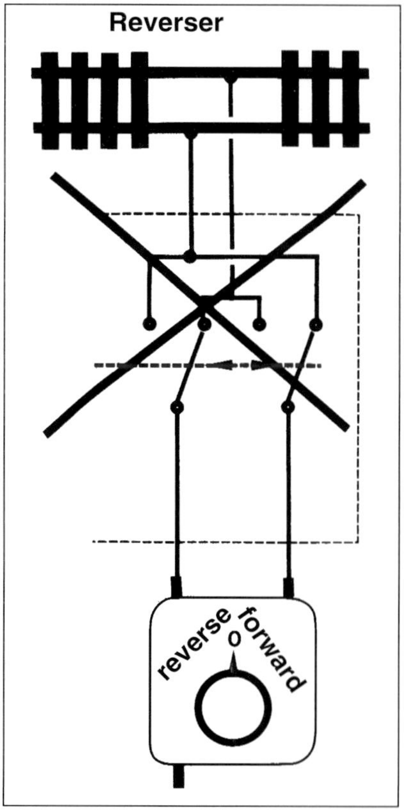

MINI-TIP

High voltage flashes with sound effects

There are often times when simulated, high voltage flashes are desirable on a model railroad layout — just consider transformer stations, arc welding in a machine shop or the flash as a trolley pole or pantograph reaches the contact wire.

In these cases irregular, intense and abrupt lighting is needed. This circuit was designed to fit this need. Connected to miniature lamps, the circuit can even simulate the flashing lights of emergency vehicles. Experiment with the outputs (1-8) on the ICs in modules A and C for variations in lighting effect.

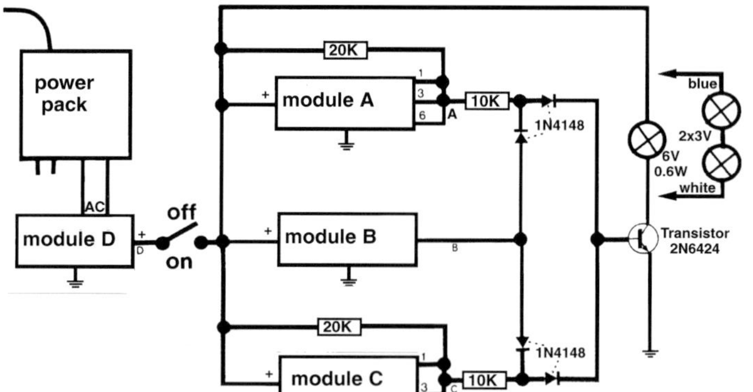

Details on page 87.

A small, sound effect generator is also include. Sound can be varied with two trim potentiometers. The generator can produce welding sounds or the sound of thunder, depending upon adjustments made. It is also synchronized with lighting effects. One of the figures shows each module individually. They take little space and can easily be built on a common circuit board. In the other figure, connections between the modules are shown.

An appropriate lamp type can be chosen to suit the application.The flashes are more realistic when blue and white light are mixed. Try coloring the lamps with felt pens.

MIN-ITIP

Circuits using LDRs

A physical description and an explanation of how LDRs work was provided in another mini tip. Here are a few uncomplicated circuits using LDRs presented with two purposes in mind. Their simplicity should make experimentation easy - try controlling relays or small motors and varying the light source at the input. Secondly, the two circuit alternatives have direct application on the layout.

If we look first at the basic function of the "Dark out — light in" circuit, the LDR is connected between the transistor's base (b) and ground. As light intensity increases, LDR resistance decreases and the transistor (or transistors in this Darlington circuit, used for switching higher effects) eventually ceases to lead current. The lamp dims. The 100kΩ trim potentiometer is used adjust the lamp's light level in relation to general layout lighting. The series connected 10kΩ resistor insures that the transistor's base is never connected directly to plus, should the trim potentiometer be adjusted to its end position. The transistor couldn't handle that. Typical usage areas are structure illumination and flood lighting. The circuit can control two, parallel connected lamps following the specs shown in the figure.

In the other circuit, the LDR is located between the transistor's base and plus pole. Because the LDR resistance value is never zero, a protective resistor is unnecessary. Function is the opposite of the previously described circuit — as layout lighting is increased, the lamp glows stronger. In this case, the lamp could be used for special effect lighting that is to be noticeable even when the room lights are turned up. This can be an unusual effect such as passenger car lighting.

Lighted outdoors and indoors

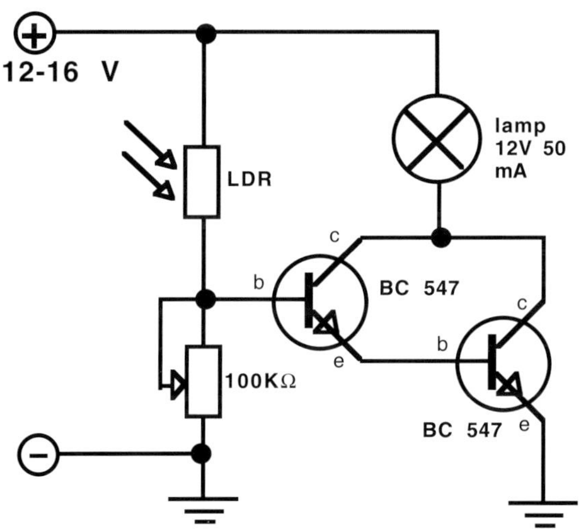

Dark outdoors – lighted indoors

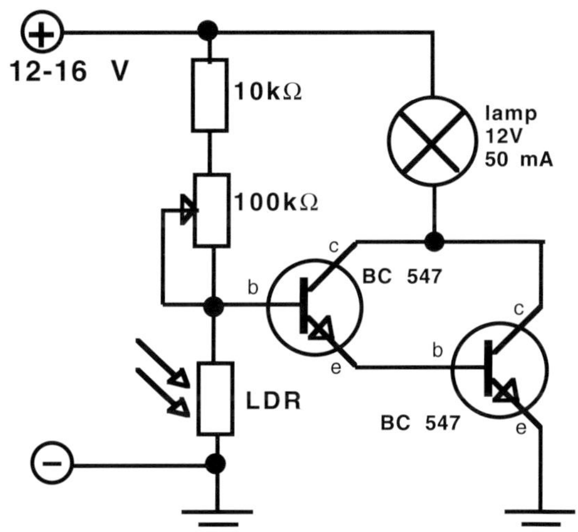

Module A & C

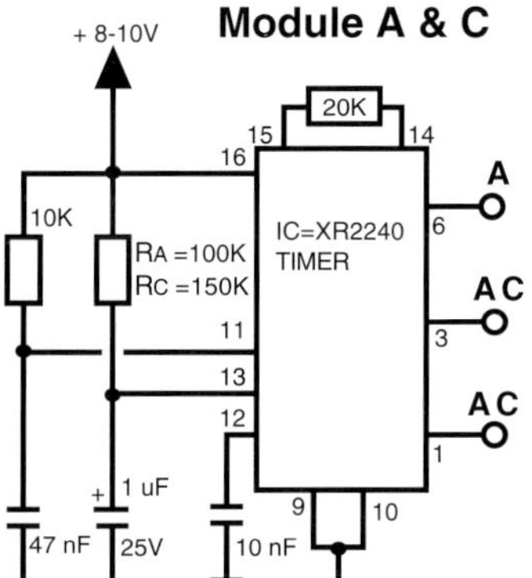

Module B

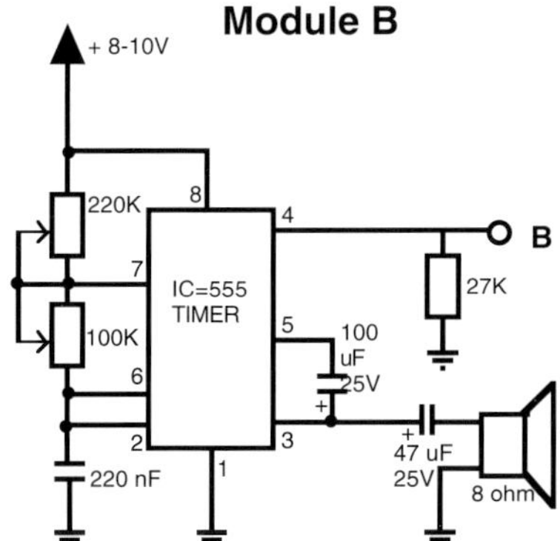

Module D

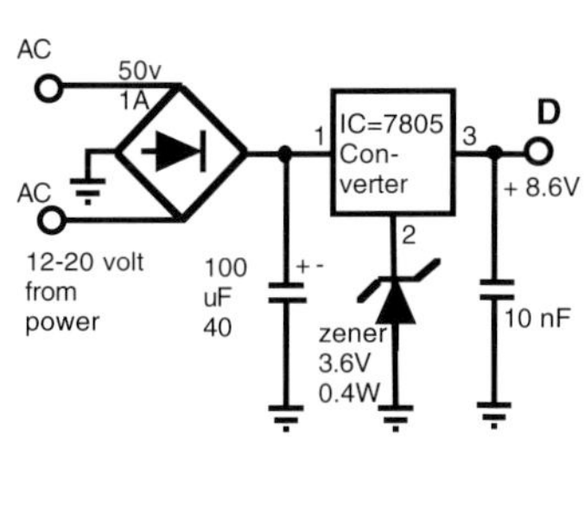

Flash unit

MINI-TIP

Measuring LDRs

Light dependent resistors, or LDRs, are useful little components for hobby work. Their internal resistance changes in relation to the surrounding light intensity. LDRs are inexpensive and are even available in sub-miniature format for surface mounting.

They are completely independent of polarity. Exactly as their more common cousins, resistors, LDRs are produced for different effects. For signal applications, in consideration of low layout voltages, you can usually get by with the small, low-effect versions.

LDRs can be recognized by their topside, comb-like structure. This is their light sensitive surface. LDR function can be checked, as shown in the figure, with an ohm meter. At a distance of approximately 10 inches from a common, 60W lamp the circuit should give a low resistance reading - typically from 100Ω-1000Ω. In darker lighting conditions, the LDR takes on the characteristics of an isolator with an internal resistance of over 1MΩ.

A common application area in model railroading is for rate regulation in sound circuits — circuits that simulate steam locomotive sound effects, for example, where the "chug rate" is controlled by motor voltage. A small lamp or LED is connected parallel to the locomotive motor and directed towards an LDR in a light-tight enclosure. The faster the train runs, the stronger the lamp or LED shines. This in turn, increases resistance in the LDR controlling (via an IC 555) the rate of sound.

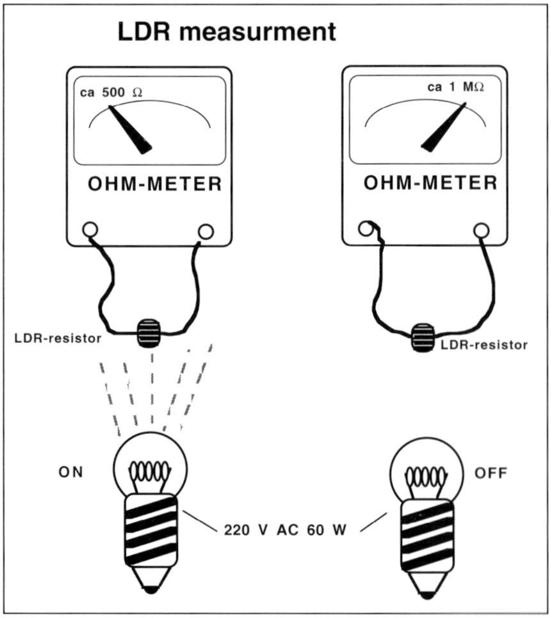

MINI-TIP

CMOS technique:

What do I do with surplus CMOS gates?

Often, we don't use all of the gates in an IC. A 4049 contains six inverter gates and the common 4001 and 4011 each have four gates. It's quite common that one or more of these gates is unused when we have finished assembling a project. If we were using the TTL circuits of the 7400 series, this wouldn't be a problem - there is no harm in leaving TTL gates unconnected. In the case of CMOS circuits ALL inputs must be connected to a logic level determined by the power feed voltage. This means that if a circuit is driven with a voltage between ground and +5V, the surplus inputs cannot be connected to 12V or any other voltage outside of this 0-5 volt "frame". This is also true for negative voltages. As mentioned in the figure text, it makes no difference if inputs are connected to high level or low level.

It is often practical to let the gate closest to the plus or minus pins be the surplus gate — it eases connection to the selected potential. Unused inputs can also be connected directly to ground without resistance. But if an input is temporarily connected to a signal source, it must be equipped with a resistor (typically 10-50kΩ) as shown in the figure. Note that up to now, we have only discussed INPUTS. An extra gate OUTPUT should be left unconnected.

It is often worthwhile to make an extra check that all inputs are logic bound. The IC will not break if an input is left unconnected but it will be extremely slow and use much current. Extremely low current drain is otherwise a characteristic of CMOS circuits.

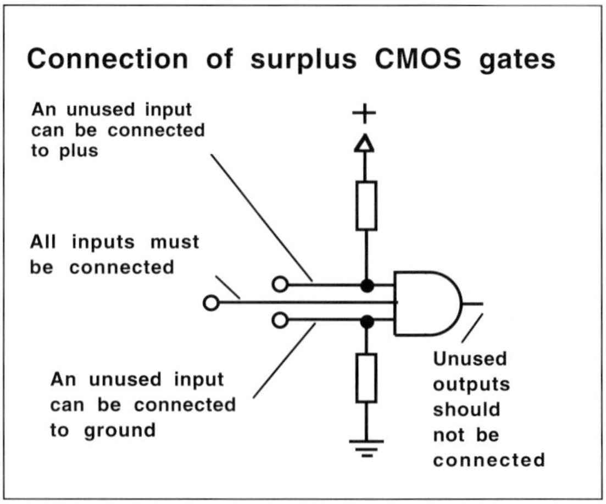

MINI TIP

Remote controlled passenger car constant lighting

This lighting module is connected between the power pack and the track, and can be used on both DC and AC systems. The figure shows connections to a Märklin power pack. For DC power packs, connect the module to the AC accessory terminals.

The two ICs use inverse feed back and provide an illuminating effect of just over 4.5W. Note that a cooling flange is used to dissipate heat. If 50mA lamps are used for car lighting, up to six may be used with the capacitor shown at C. Without the capacitor, light intensity is weak at startup and increase with train speed. When cars are in a trackage block where power is cut, constant lighting is maintained by a capacitor, D, at the block's rail gap.

This high frequency lighting generator has one component that we make ourselves. It's the coil. Although I used a ferrite rod antenna from a scrapped transistor radio as a coil core, ferrite rods can be found in electronic supply stores. I wound the rod with approximately 100 turns of 0.6mm, lacquered copper wire as a primary winding. The total number of secondary winding turns over the primary winding varies depending up power feed voltage. At a feed voltage of 14-16V, 20 turns is just right for 3V lamps. If you increase this to 90 turns, you can use 12V lamps. As you can imagine, a little experimentation is in order.

Lighting can be turned on and off with the module's switch without effecting other functions. A 100kΩ trim potentiometer adjusts light intensity.

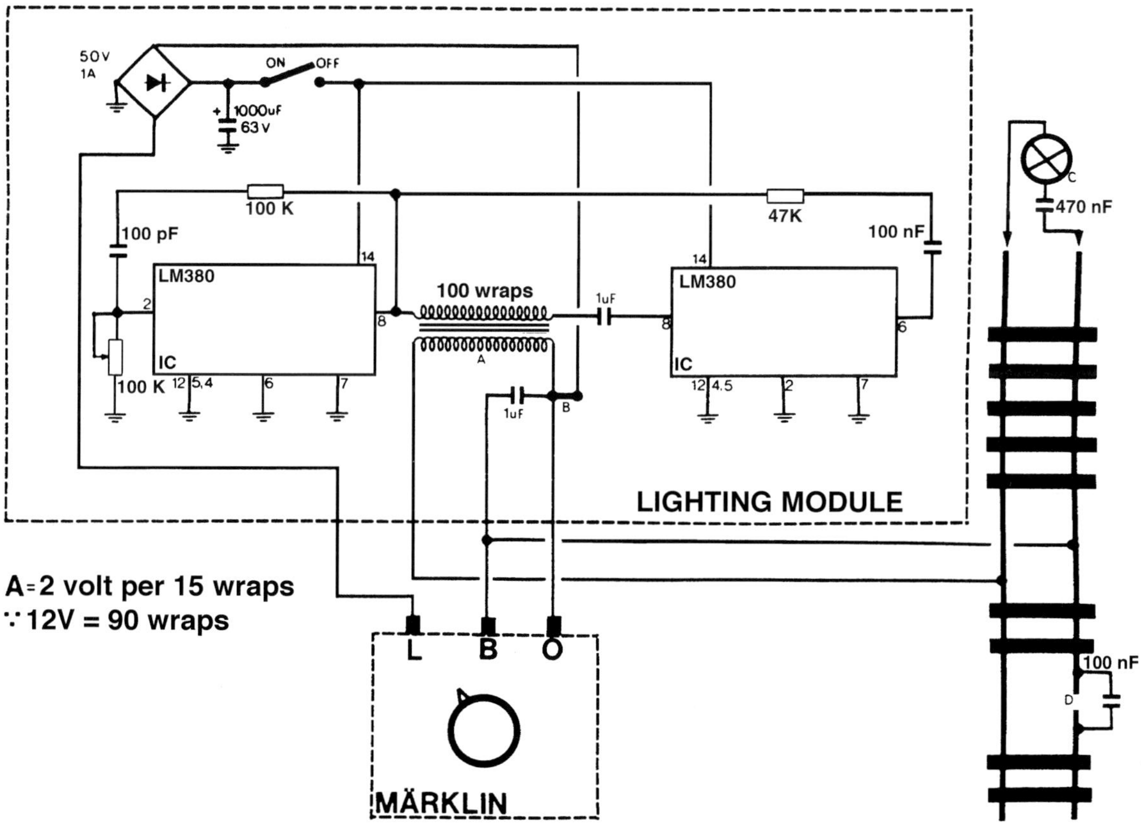

MINI-TIP

CMOS- Technology:

Different ways of connecting LEDs

The LED is a light emitting semiconductor that is polarity dependent. The polarity must be correct for the LED to light. This applies to all LED colors and sizes.

A general rule is that an LED should always have a current reducing series resistor (or other more sophisticated current limiter). We will now take a closer look at this general rule.

If we use an unbuffered IC, such as a 4081 (containing four, double input AND gates), an LED can be connected directly to an output, providing that the power feed is in the 3-9 volt range. Note that all LEDs can be made to light with either high or low logic levels, depending upon how they are connected.

When connected to higher voltages or when used with ICs of a type other than standard, buffered ICs, a series resistor of approximately 680Ω-1kΩ is connected as shown. An IC designed for lighting LEDs is 4511, a 7 segment display driver. A display, in this context is a circuit with seven LEDs mounted in an "eight" that is used to display all numbers and letters from a-f. Series resistors should be connected to the 4511's outputs.

If higher loads, such as light bulbs and multiple LEDs, or a relay shall be regulated by a CMOS gate, a "servo" (consisting of transistor circuitry, for example) is required.

Gates as LED drivers

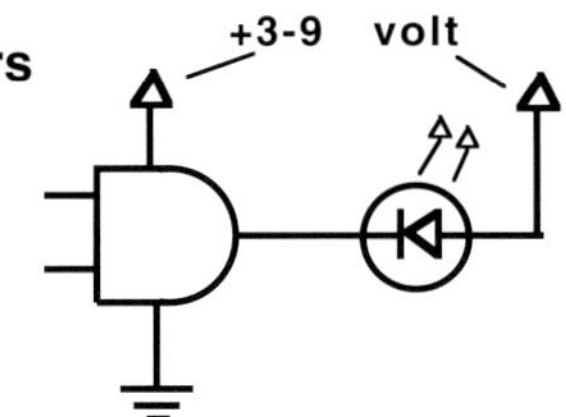

LED on when gate output is low

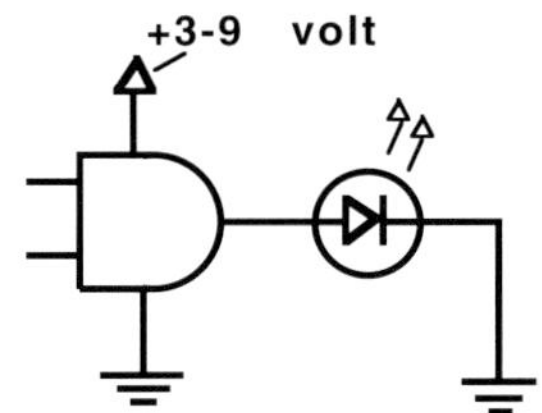

LED on when gate output is high

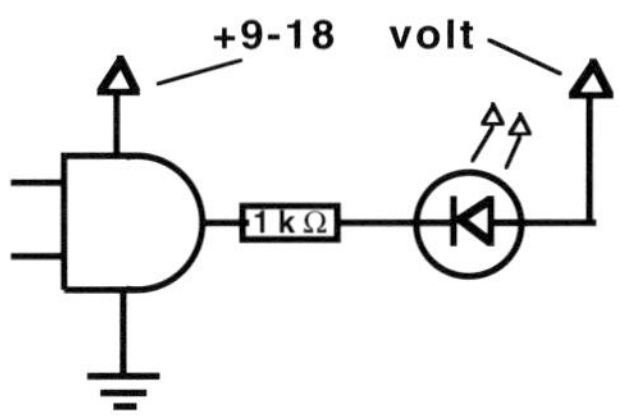

LED on when gate output is low

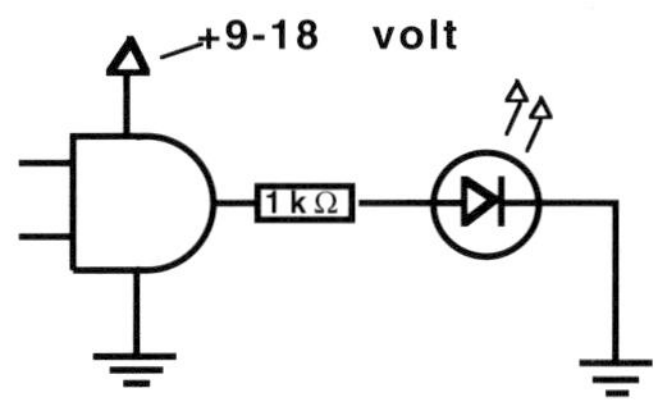

LED on when gate output is high

MINI-TIP

The ABC of the LED

(Light emitting diode)

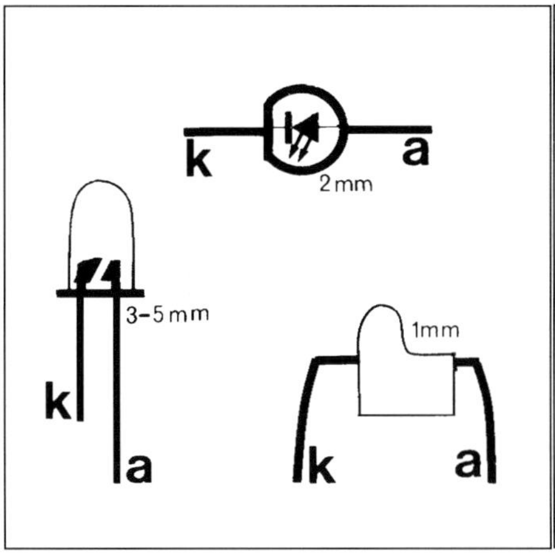

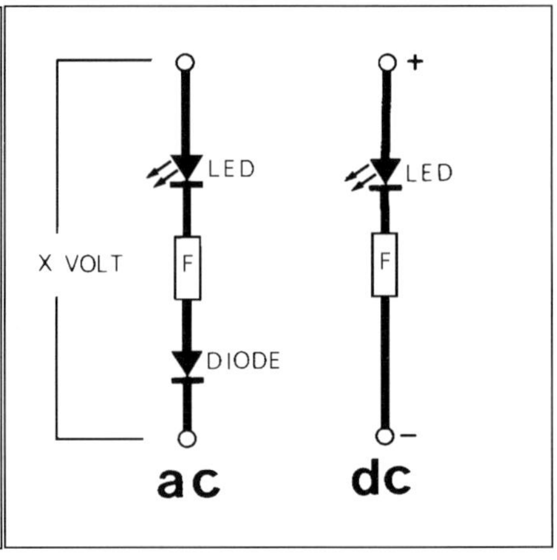

MINI-TIP

CMOS Technology:

Slow and quick functions

There are a few simple guidelines to remember for attaining maximum performance from CMOS gates. The figures show examples of these. Long leads increase capacitive effect (or "capacitor" effect). This puts a heavy load on the gates. The gates react slower in switching between logic levels and can become a "bottleneck". Another problem can occur when gates with several inputs are mixed with gates having few inputs.

A gate in the standby state has a static effect consumption that is extremely small. An inverter at 5V draws just 100-200pW (a picowatt is equal to 0.000000000001W). A gate that is operating (switching between ones and zeros) has a dynamic effect consumption measured in milliwatts. This considerably higher power consumption is of a very short duration however, allowing the CMOS to be powered for long periods of time with miniature batteries for memory retention.

Input load at gates

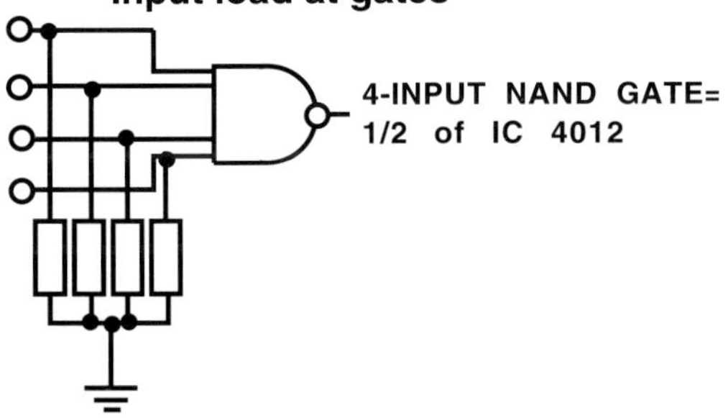

Several inputs = slow operation!

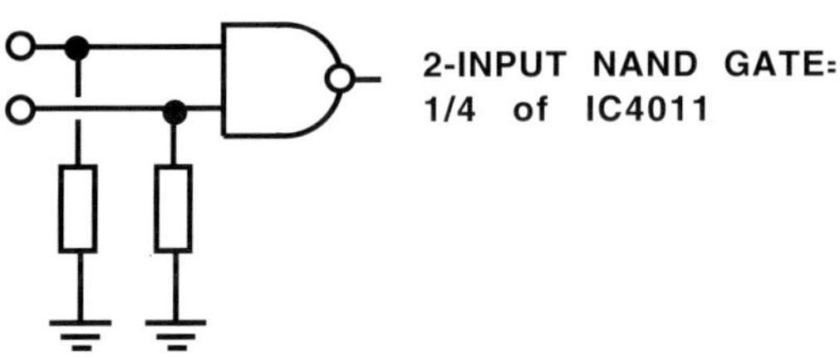

Few inputs = quick operation

Capacitive load on gates

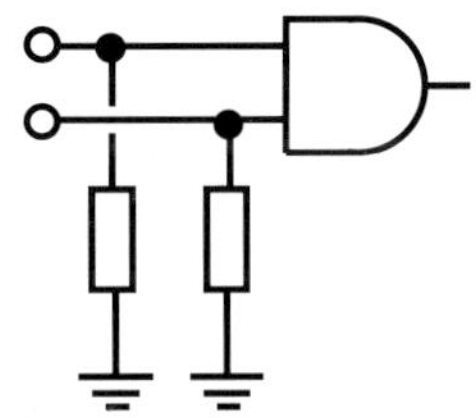

Short wires = quick operation!

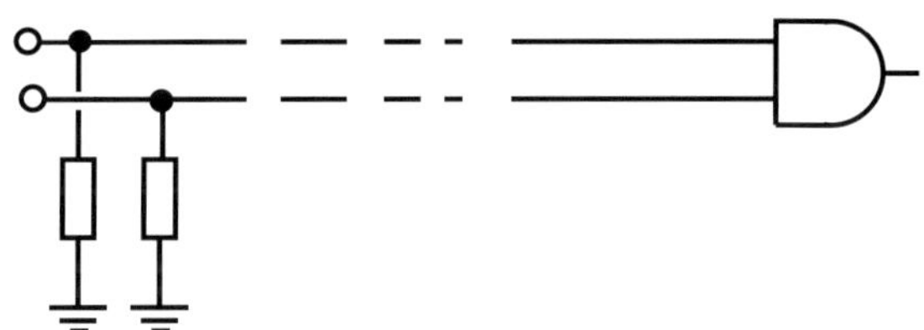

Long wires = slow operation

SERIES RESISTORS

volt	DC	AC
6	680	330
12	1 K	470
18	1.8 K	820
24	2.2 K	1 K
30	2.7 K	1.5 K
40	3.9 K	2 K

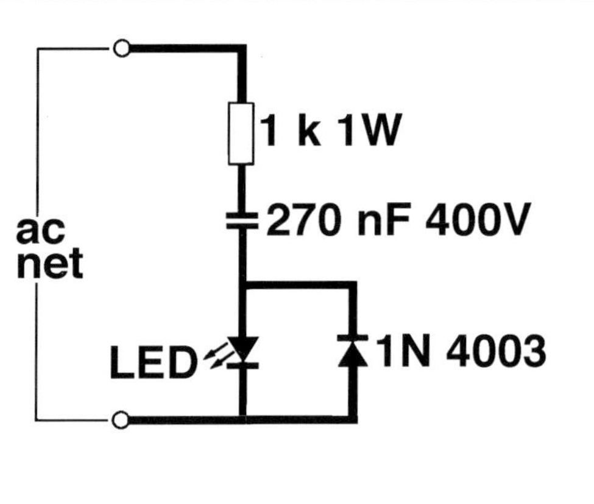

Polarity detector for DC operations

When the layout has grown to the point that trackage is divided into blocks and powered from different controllers, complex situations arise. Wiring becomes more complicated and a certain amount of planning is required. In this mini tip we will concentrate on trackage blocks, or to be more exact, the gap between blocks.

All too often, trains running from one block to the next are stopped cold. This often happens when polarity in the second block does not match polarity in the first — plus rail meet minus rail and vice versa. Although this is easy to remedy — just reverse polarity in the second block — aesthetically, it's not a pretty sight when trains hit this invisible barrier. It's not good from the electrical standpoint either. The locomotive wheels bridge the gap separating the blocks and this causes a short circuit. Detectors indicating mismatched block polarity are useful for preventing this all too common occurrence.

A similar case that isn't as serious, but nearly as irritating, arises when power to the next block is either shut down or is too low. A detector is useful for avoiding these annoying incidents too.

This type of problem is also common in turning loops and on double track, passing switch trackage.

The detector is easy to build and simple to adjust. The load circuitry, with four resistors at the block gap, has appeared in the American model press several times and in various versions. The beauty of this circuit is that it can, despite its simplicity, detect running voltages at different levels and polarities, and is also free from a common ground reference plane. The only connections are via the four diodes. These are followed by two resistors. These could, of course, be substituted by a single 4.7kΩ resistor but it is appealing to offer constructions using identical component values. The capacitor is needed only with pulse power controllers. The trim potentiometer is used to set the track voltage, triggering level for polarity indication. Indication can be via a control panel mounted warning light or trackside signal, as shown in the figure. The connection points (S, plus and ground) on both units should be connected together. Although I have used two signal masts, you can use one if you wish. A detector is needed for each block division point.

Note that each detector uses just one of the four operations amplifiers contained in the LM324. Because it is not a CMOS component, unused pins may be left unconnected. If you prefer reversed signal indication, switch the connections at IC inputs 5 and 6.

In passing, this and the other circuits presented in this book were not designed with command control in mind. If you use command control, system compatibility must be considered before connecting these circuits to the layout.

Polarity detector

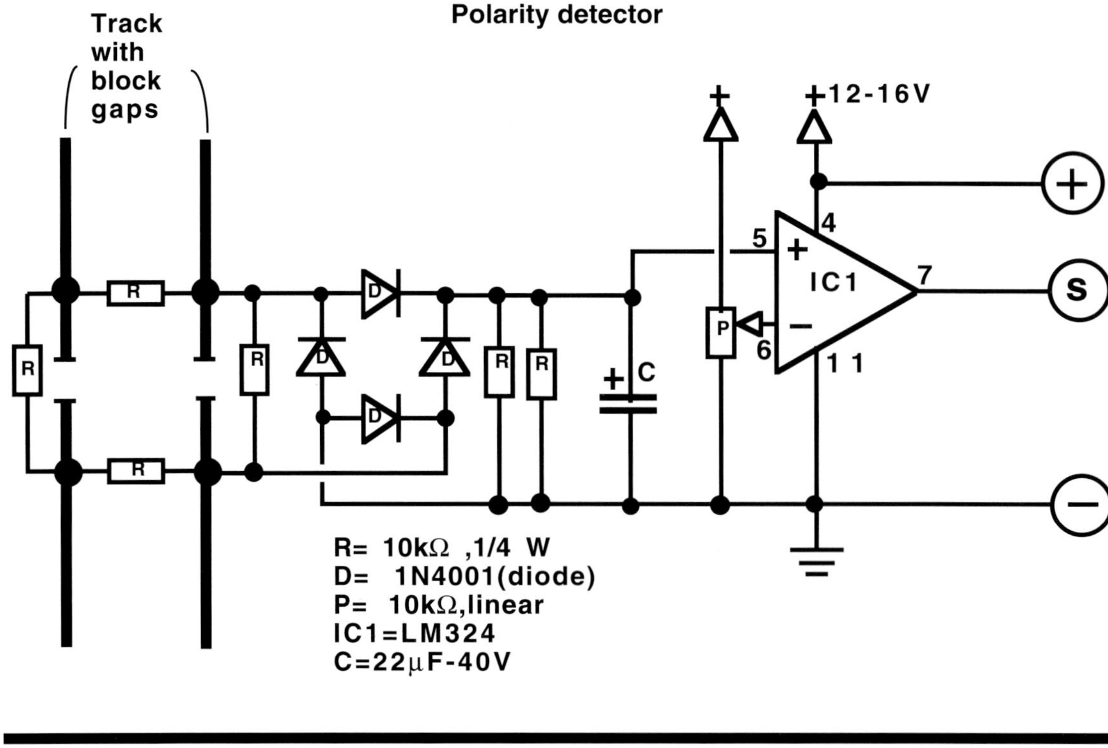

Linear IC LM324
4 operation amplifiers

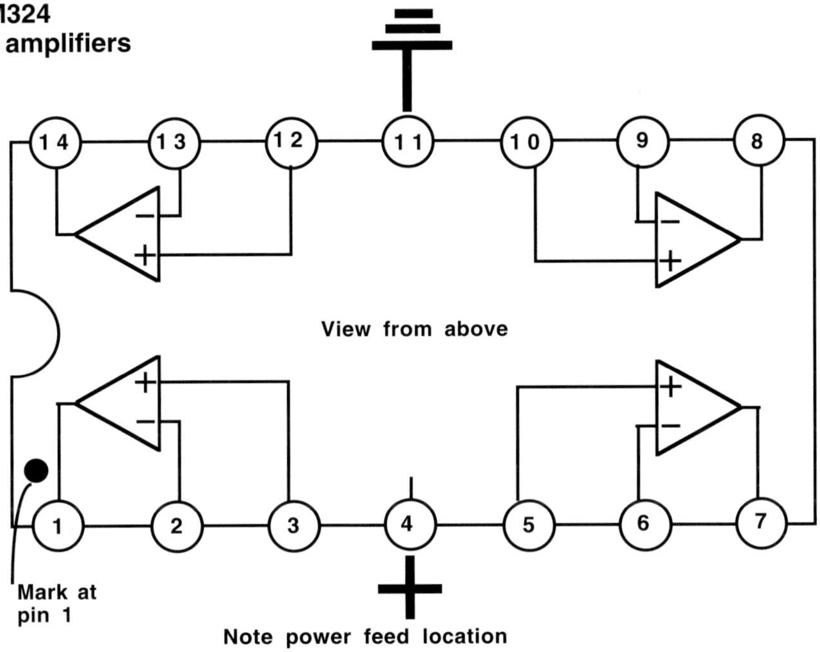

Signal masts

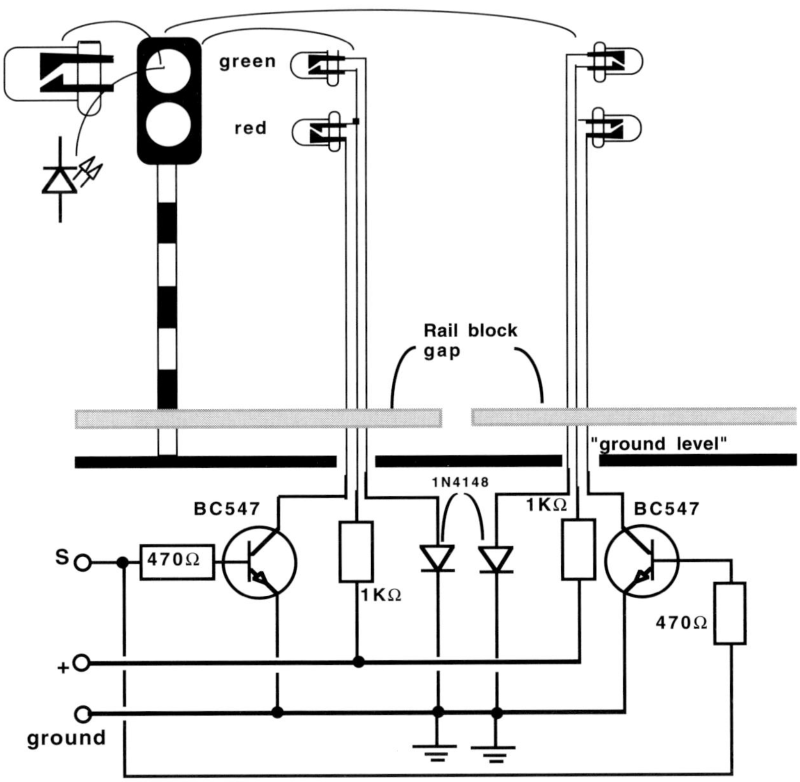

Reverser for an AC-locomotive

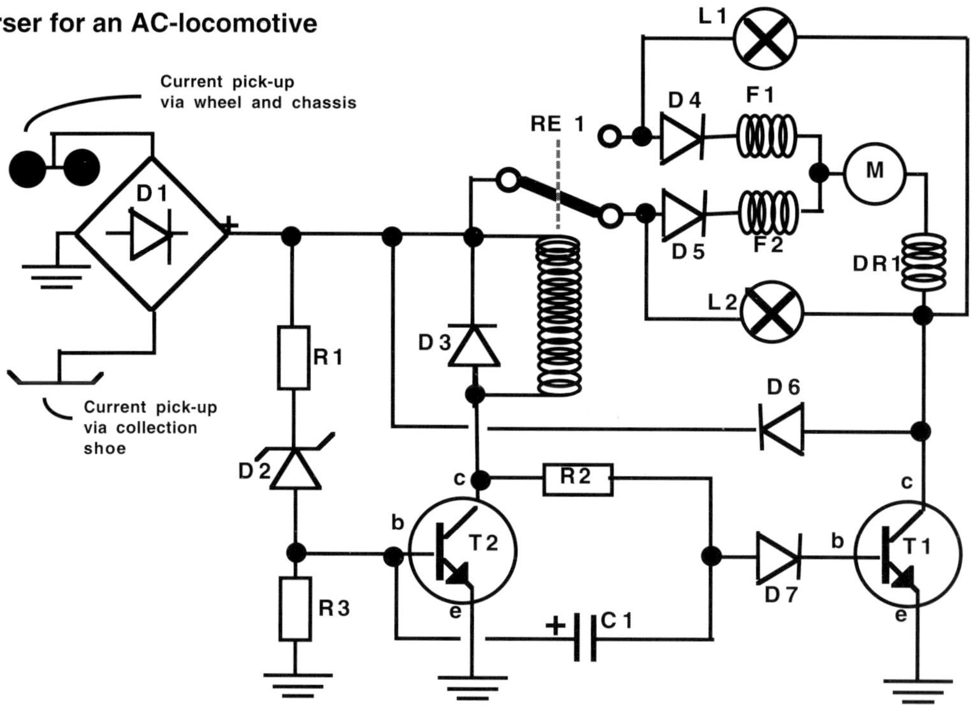

Here are some commercially built electronic direction modules . The top left is black because it is totally isolated via a rubber coating. It has the advantage of not risking any unintended short to the engine chassi, but it is not very lightly that you ever want to try to replace a faulty component on it.

The second one to the right is too big for many engines, but it is fairly easy to make repairs on and the components are standard ones. Then finally the smallest one with a silver button battery onboard. The disadvantage is change of battery even if that doesn't occur very often. But it keeps remembering last made direction command which is very handy if having several track sections with power off selector in front of home signals. The components on this module are mostly surface mounted (SMD).

MINI TIP
For AC locomotives:
An electronic reverser

Märklin manufactures robust and dependable locomotives for AC operations. The reversing switch is an onboard, electromagnetic, bistable relay with two field coils (for forward and reverse). Märklin locomotives often change direction of travel with a hop and a bright flash from the headlights.

This electronic complement eliminates these disadvantages without altering reversing functions. It is a variation of the Märklin construction and has the benefit of employing inexpensive, standard components and the original relay. The component list is divided between existing components and those that must be added. Note that older locomotives were made differently and that this alteration works best on larger locomotive with plenty of space under the shell for the new components.

As can be seen in the schematic, current from the wheels is lead through the frame, to a rectifier, D1. Because this point is normally connected to relay RE 1 (the common contact) and the lamps, it must be insulated. The original lamps are replaced with 20V/50mA lamps with two connection leads and the relay is unscrewed and re-mounted using glue and an insulating washer.

During normal running, motor current is lead through T1 which receives base current via RE 1 and R2. When

MINITIPSET

CMOS TECHNIQUE:

When working with high frequencies...

CMOS technique is in many ways ideal and is quite flexible. We have previously mentioned that it can be powered with DC within a large voltage range — from 3 volts up to a maximum of 18 volts. It should be mentioned that a few manufacturers have set 15 volts as the upper limit. This tip will cover CMOS technique's dependency on relatively high voltages for work with very high frequencies. A typical value at 5 volts is about 2.5Mhz (one megahertz is equal to one million hertz). This means that we can connect to a maximum frequency of 2.5Mhz (an indication of how fast we can switch between high and low level states).

If the voltage is increased to the 12 volts, the frequency limit is twice as high. This means that the gates can work with frequencies of up to 5Mhz. To raise the voltage to the maximum limit will provide only marginal increases in frequency.

It should also be pointed out that unbuffered ICs (A series, 4011A for example) have a less distinct transition between logic levels as compared to the buffered version (B series, 4011B for example). The most important difference in everyday use between these two is that the buffered ICs have protective diodes at the inputs to reduce the effects of static electricity. The markings on the ICs specify whether they are buffered or not.

Voltage and frequency relationship

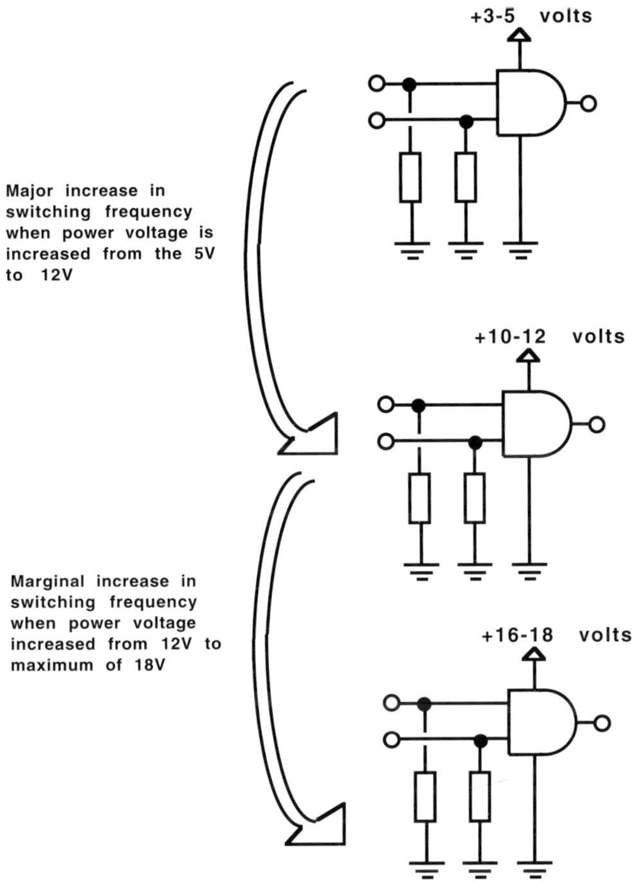

reversing, a momentary high voltage, bridges zener diode D2 and "switches on" T2 after the electrolytic condenser C1 has "switched off" T1. This function order allows us to avoid the otherwise, obligatory hop when changing directions.

Additional components

D1 Diode rectifier 40V-1 A
D2 Zener diode 24V-400mA
D3 Diode 1N4148
D4 Diode 1N4001
D5 Diode 1N4001
D6 Diode 1N4001
D7 Diode 1N4148
R1 1KΩ 1/4 W
R2 470Ω 1/4 W
R3 10KΩ 1/4 W
C1 10µF 25 V
T1 TIP 41 A
T2 BC337

Original components

RE 1 Reverser relay
L1 Front light
L2 Field winding - forward
F2 Field winding - reverse
DR 1 Choice coil
M Motor

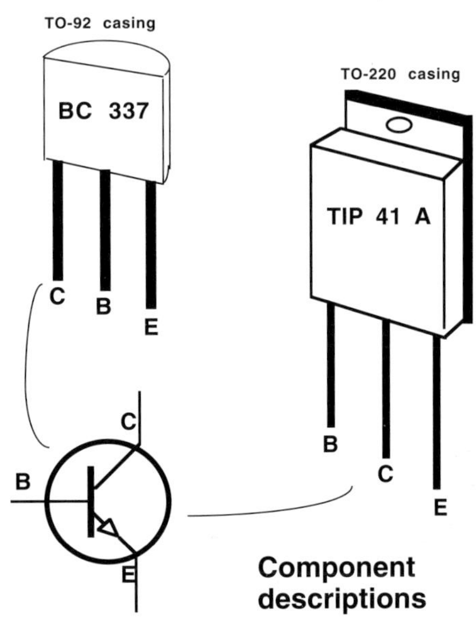

Component descriptions

MINI TIP

Time relay for intermittent operations

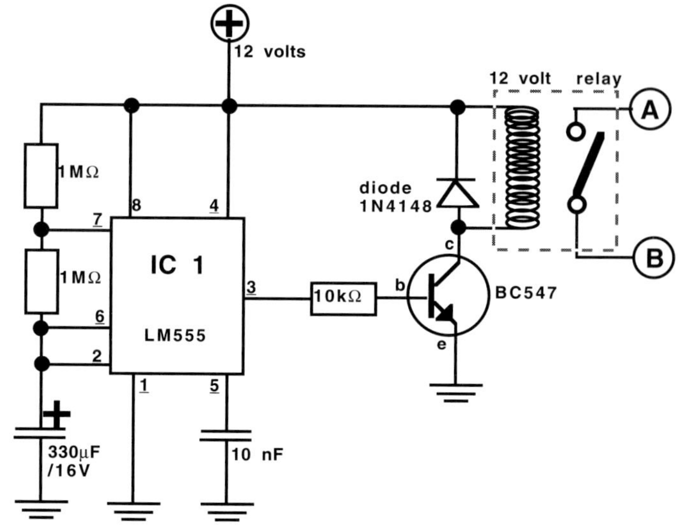

Model railroads often have somewhat odd, electrically powered operations that one would like to run periodically but that the railroad's head dispatcher just doesn't have the time to attend to. Here are a few such examples. A streetcar on a city line runs on a loop and makes periodic stops. A funicular railway makes a short run and then stands still for a while. Or why not a train, running on its own loop?

Here is an extremely simple controller for these applications. Using the suggested component values, it provides an eight minute run and a four minute stop in an endless cycle. If the upper 1MΩ resistance is reduced, the total cycle time is reduced at the cost of run time and nears a fifty-fifty ratio.

If the lower resistance is lowered in a corresponding manner, the total cycle time is reduced but the run/stop ratio is not effected.

For Z scale, with its 5V operating voltage, this circuit can almost be used as is. The LM555 can be powered with as little as 3V but the relay should naturally be replaced with a type for lower voltages. Points A and B are series connected with the system to be controlled, just as one would connect a common on/off switch.

MINI TIP

End of train device with constant lighting

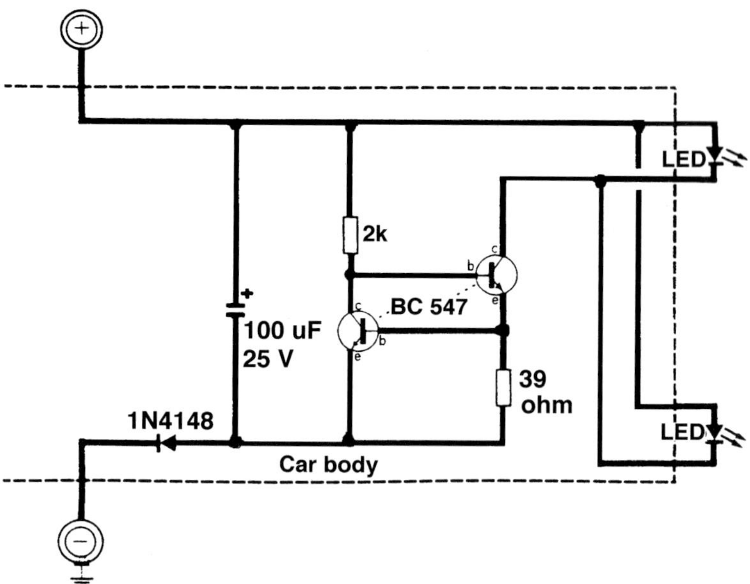

These eight components can be bought for a few dollars at an electronics store. The LEDs can vary in size from 1 to 5 millimeters in diameter and the circuit is suitable for most scales. With DC, the circuit is directional while on AC layouts, the LEDs will light in both directions of travel.

The transistor circuit functions as a current limiter and provides a constant lighting intensity from about 2.5V up to maximum running voltage. Because locomotives normally start at a voltage of 3-4V, the device will light even while the train is stopped at a station.

Lighting intensity is varied by exchanging the 39Ω resistor with resistors of other values. For the compact constructions required in the smaller scales, solder the components directly to each other, omitting the circuit board. In these cases, begin by gluing the two transistors to each other. Follow by soldering the other components directly to the transistor leads and insulate as required.

If the car doesn't already have interior lighting, a method of current collection must be found. For the Fleischmann car shown in the photo, truck-mounted contacts are available for current collection.

MINI-TIP

The red caboose lights

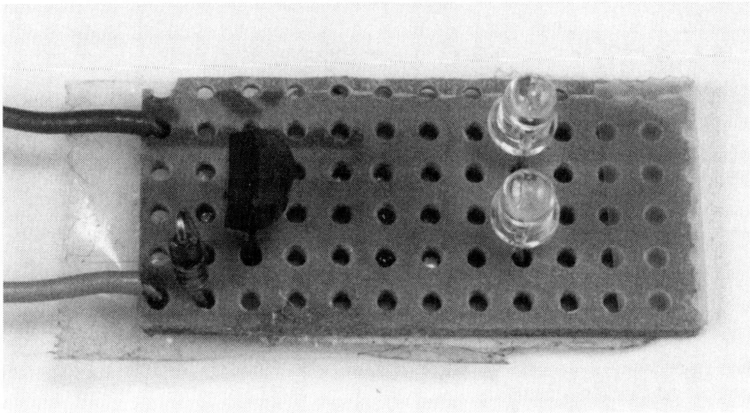

The classic analog lighting problem is to get light to be "on" while train is standing still. And added to that you normally hardly see the lights while the train is idling or at low speed, but superglowing while running close to full speed.

There are a number of ways to deal with the problem and here is one of those simple ways to get around it. We will use just 4 components to achieve this; two LED´s, one diode and one J-FET transistor (field effect transistor of N-channel type).

The LED´s are our red lighting sources and they will get correct current limitation through the shunt acting semiconductor. By introducing the diode in series with the rest of the components we define the direction of the current, regardless of polarity of the track feeding for the moment.

If we wanted just constant light anytime, we would substitute the diode with a rectifying bridge. Here we do want it to recognise the direction, since the caboose lights only should be red towards the rear. There is a small sketch showing that function. Do note that this project works for both DC layouts as well as Märklin ones.

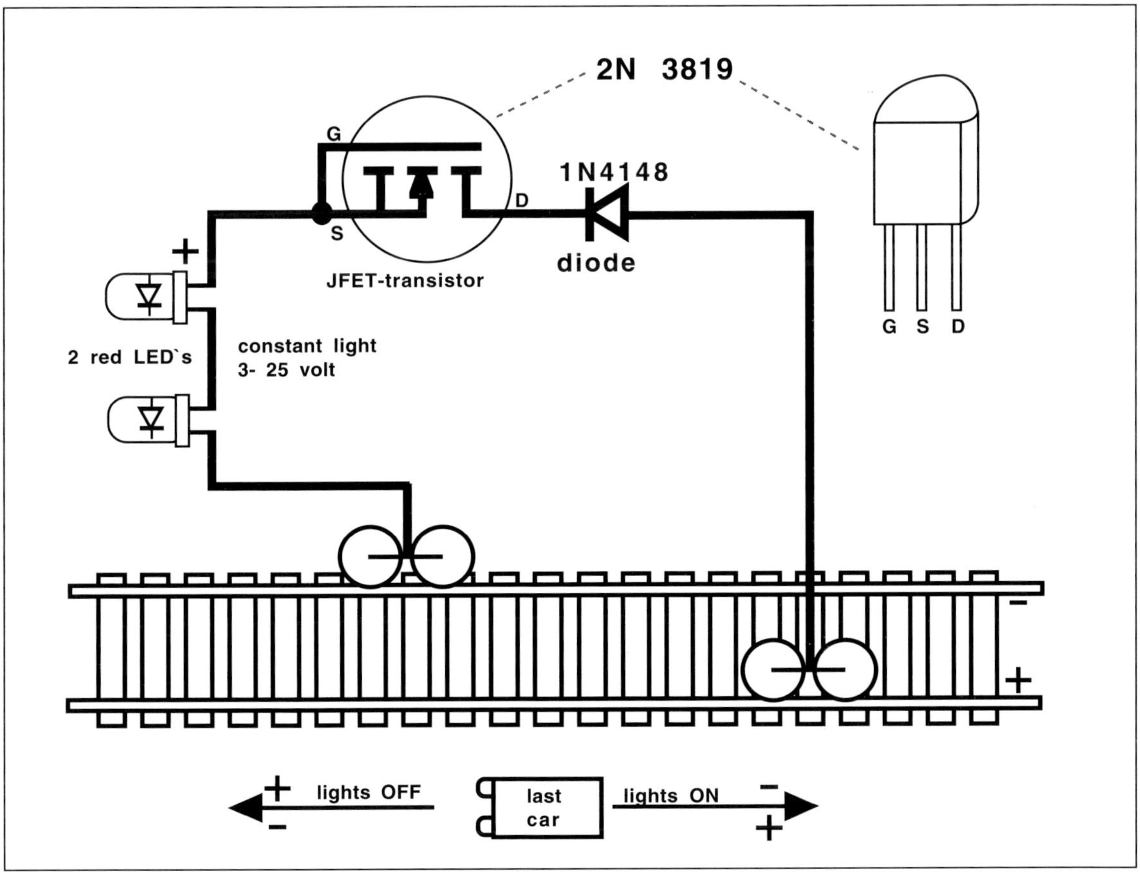

MINI-TIP

Flash generator for Railroad X-ings

This mini-tip actually comes in two versions; one basic and very easy to build with a minimum of components; and one with some more sophisticated adjustment. Let's look at the basic one first. It uses the classic and priceworthy timer IC 555 and in this design I have given it a fix flash ratio being as close as possible to the true flashing frequency of a typical domestic US X-ing signal.

There is a small design feature beyond class one in electronics. The shape of the square pulse is improved by connecting the 680 kohm resistor to the output pin No. 3.

Flash generator for road x-ing signal

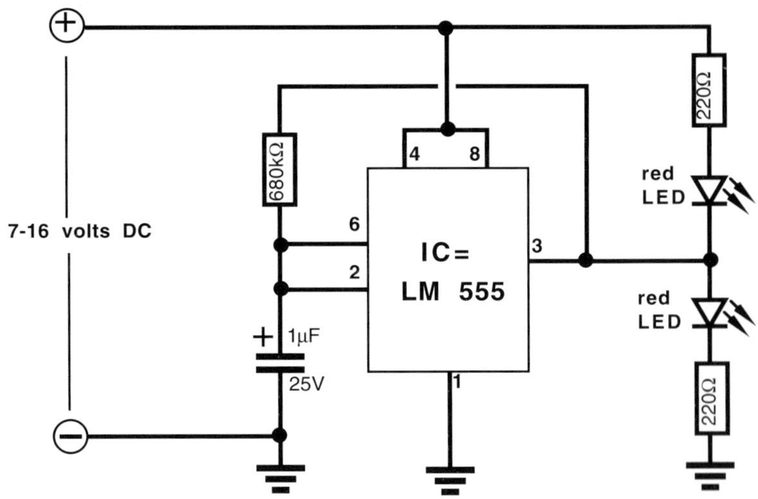

Then we have a simple but effective output section making use of the muscle found on the pin 3. It can both sink one LED light to ground and in the next sequence of the output pulsing feed the other LED and its current limiting resistor (220 ohms) with power plus. And that is the total circuit.

Our deluxe design has an identical output section as the basic one, but has a few components added to the timing section in order to generate a perfect 50-50 ratio in the flashing operation. The main flashing frequency is still generated by the 680 kohms resistor and the 1 microfarad electrolytic capacitor, but the balance adjustment which is equal from an electronic point of view to charging and discharging the capacitor, is substancially improved. Most often it is quite sufficient to make the adjustment on the trimmer while watching the flashing operation, and no instrument should be needed for that operation. But if you do want a perfect pulse, do use an oscilloscope and connect it to pin 3 and pin 1 (ground) of the IC.

Flash generator for road x-ing signal – with pulse adjustment

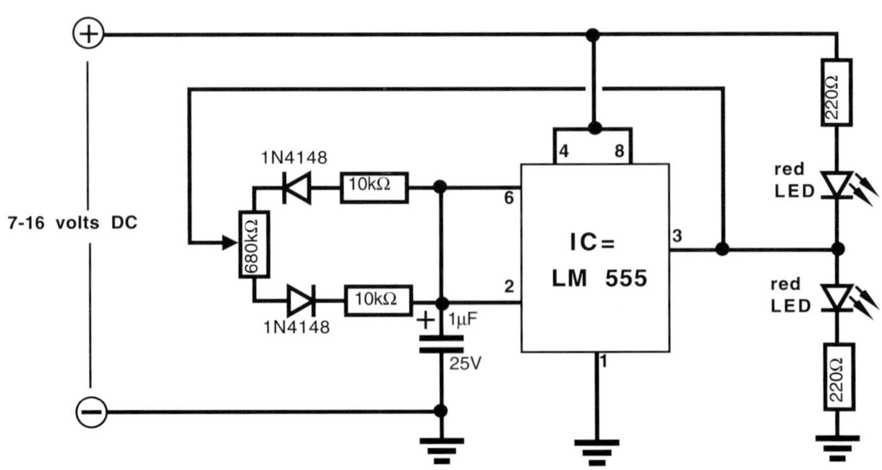

These are the most common model railroad scales globally. The Miniclub from Märklin (1:220) can easily operate inside your drawer or on a small tabletop, but there is not much space inside for electronic modules. Still I have been successful in mounting full decoder equipment inside 3 locos. N-gauge (1:160) is more space friendly and will with todays compact components allow certain light projects and even sound if you can get hold of small enough loudspeakers.

H0 is the most common scale and accordingly most projects in this book fit into that scale. And finally the G-scale(1:22.5), or its close relative, the 1-scale (1:32) which allows almost any feature to be built in. If you are building in the latter scale, don't miss the opportunity to mount very large loudspeakers onboard, since the quality improves dramatically.

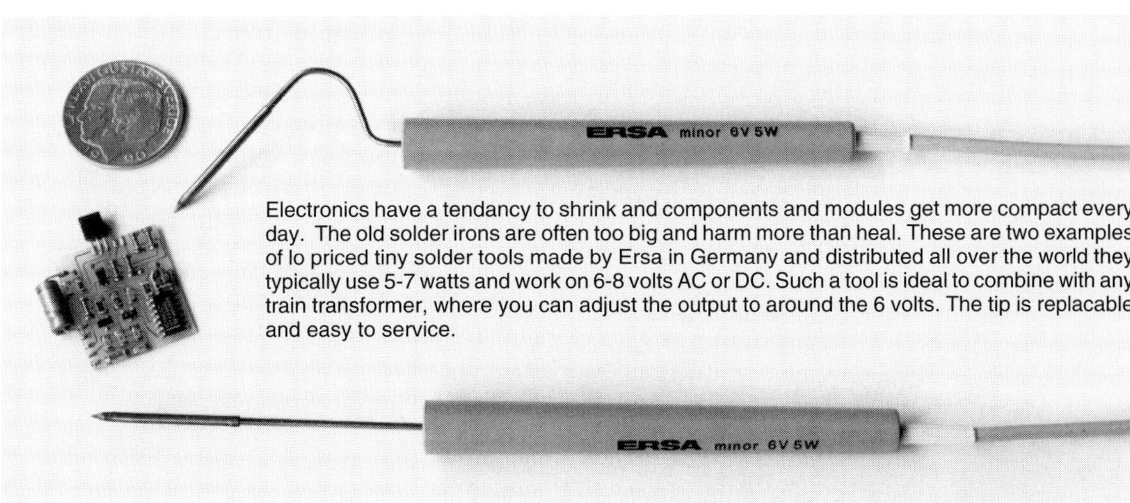

Electronics have a tendancy to shrink and components and modules get more compact every day. The old solder irons are often too big and harm more than heal. These are two examples of lo priced tiny solder tools made by Ersa in Germany and distributed all over the world they typically use 5-7 watts and work on 6-8 volts AC or DC. Such a tool is ideal to combine with any train transformer, where you can adjust the output to around the 6 volts. The tip is replacable and easy to service.

COMPONENT DESCRIPTIONS

Here and on the following pages are descriptions of most of the components used in the book's various projects. With the help of the figures, you can easily translate the schematics to connection points and soldering locations on the circuit board. As an extra aid, the figure below shows the most common electrical component symbols. The reed switch figure was included because reed switches figure so prominently in many of the projects; because it is an important component; and because one should handle them with care. Be especially careful when bending their leads — the glass casing cracks easily.

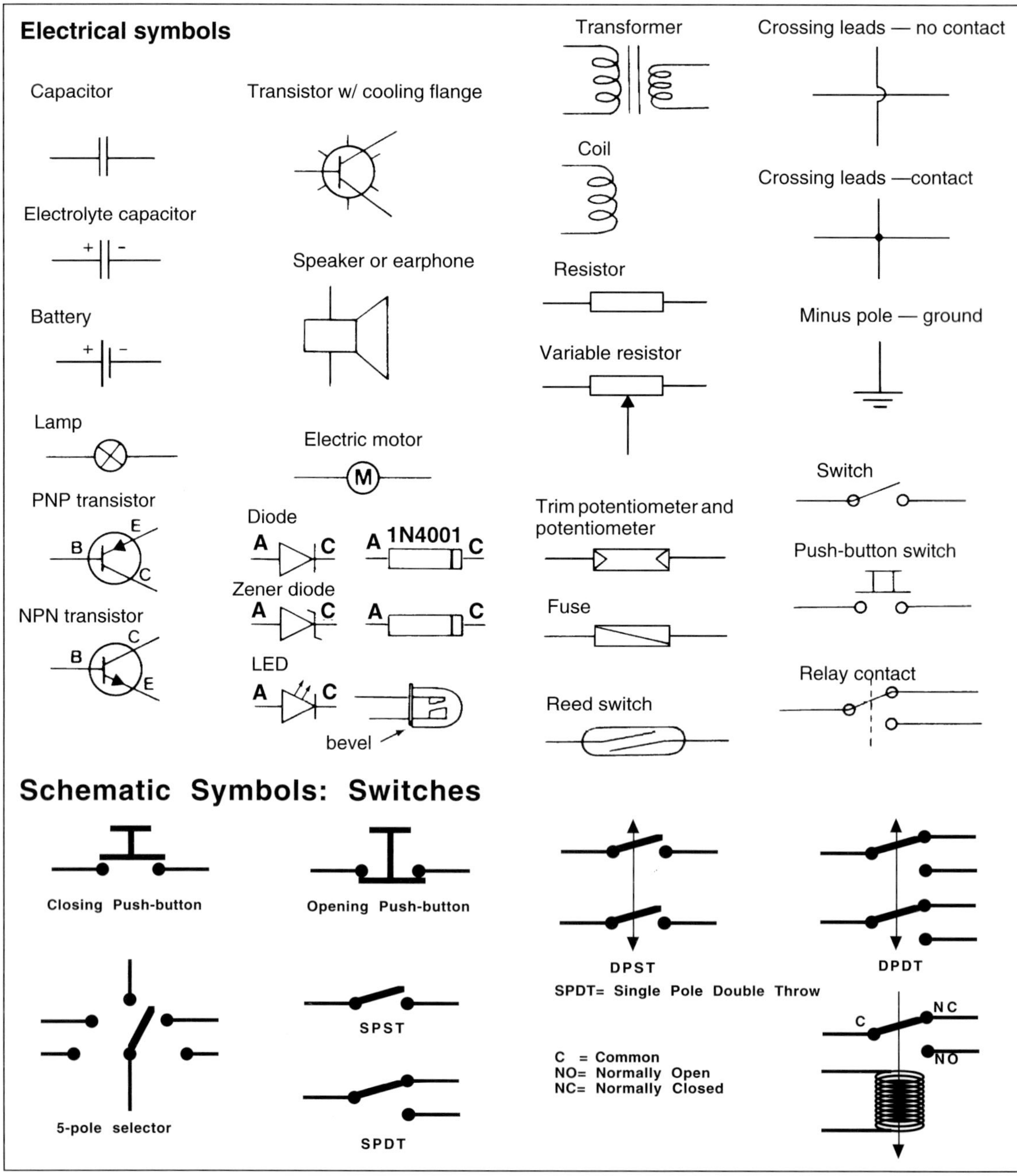

IC 4001 CMOS circuit
(4) double input NOR gates

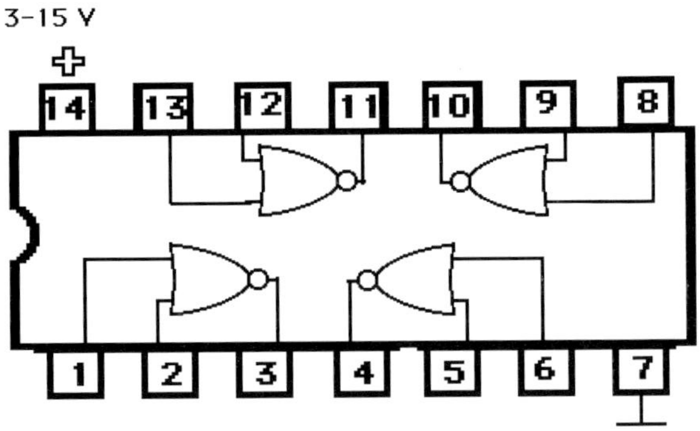

IC 4011 CMOS circuit
(4) double input NAND gates

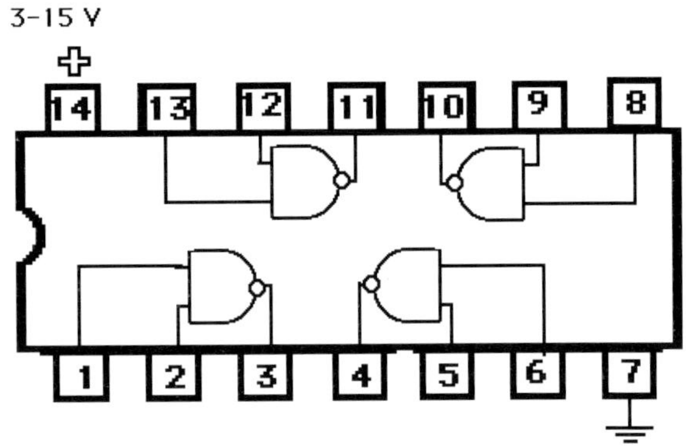

IC 4013 CMOS circuit
Double Flip-flop

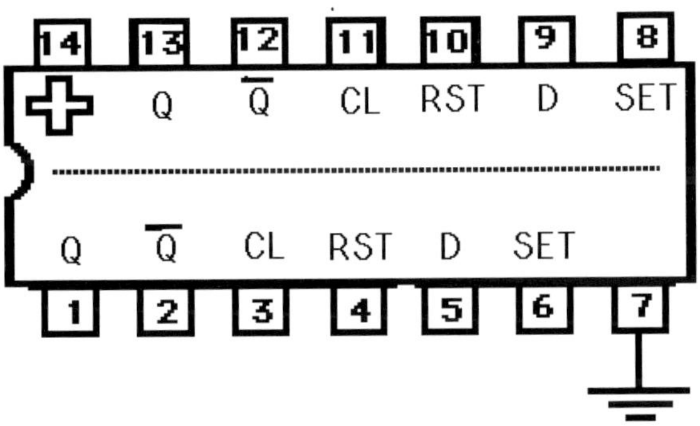

IC 4093 CMOS circuit
(4) double input NAND gates
Schmitt trigger type

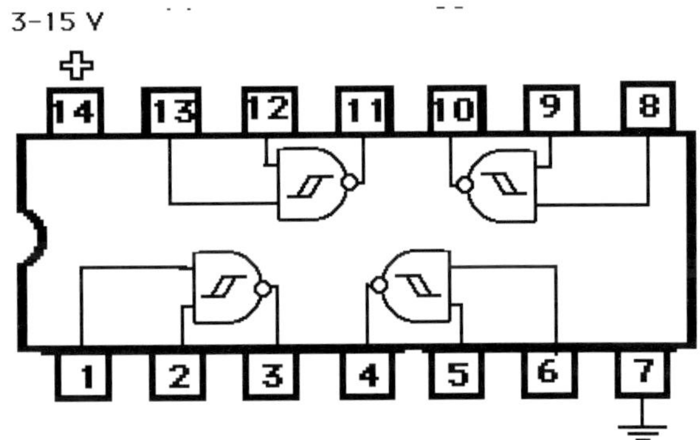

$\bar{Q}$: output that has inverted value of Q

Cross Reference - Transistor Guide	
European types	**US-types**
BC 107	2N3855
BC 327	2N4289
BC 337	2N4952
BC 547	2N3707
BC 556	2N3703
BD 135	2N4077
BD 136	2N4078
BD 679	2N4077
BD 680	2N4078

Reed switch
Contact pair closes in magnetic field

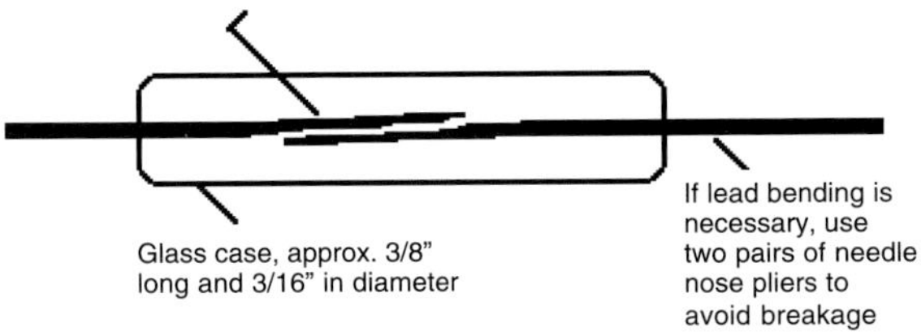

Semi-conductors in TO-92 case

Small signal transistors

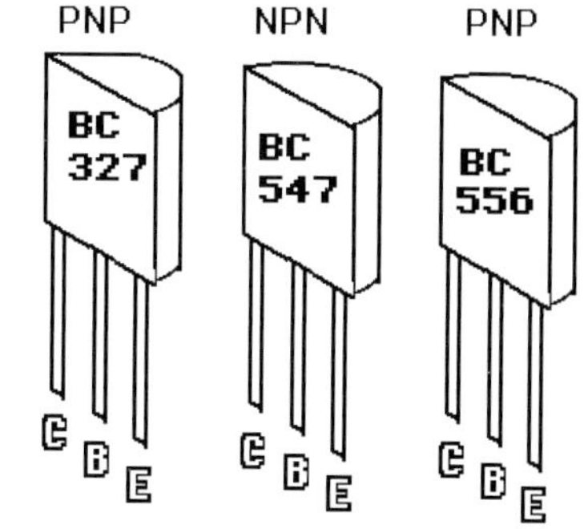

C=collector
B=base
E=emitter

Voltage stabilizer

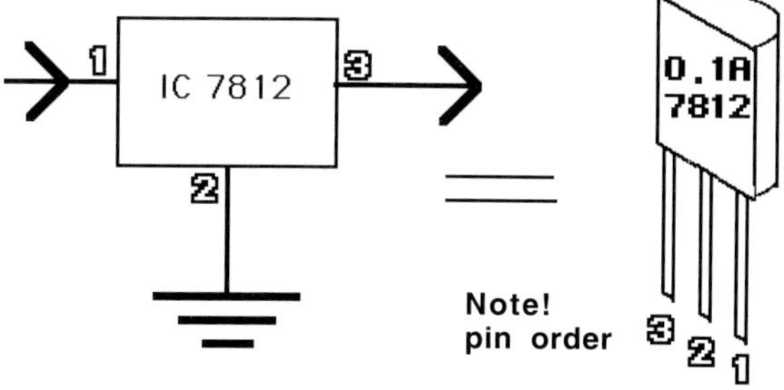

Semi-conductors in TO-220 case

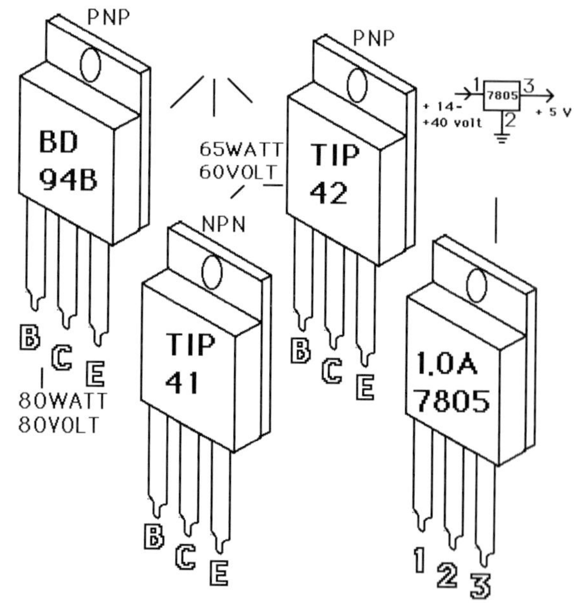

Mini-tip

How to operate Swiss Can motors

A majority of Model Railroad engines are powered by standard electric motors having an iron rotor with 3 or 5 poles. They are reliable, simple in design and low-cost to the producers. They are recognised by start-up of having an easy to define friction threshold. You increase track voltage with the knob slowly and it does take a while and a certain voltage level to overcome this friction barrier. Then suddenly the engine makes a very firm move and you tend to back off with the knob in order to get slow motion. In fact you can turn the knob off quite a bit before getting the engine to a stand still. This is the typical iron rotor electric motor, often also named "open frame" since you can watch the internal rotating operation.

A second group is named can motors. They are simply built into a small can and let Sagami as a leading brand represent these very compact motors, still having iron rotors. They tend to have lower friction threshold, but there is one still to overcome.

The third group, and in my opinion the "Rolls Royce" group, clusters the Swiss can motors with air coil rotors. To my knowledge when this is written (Sept. 96) there are 3 producers using this unique Swiss patent. They are Faulhaber, Escap and Maxxon. From outside they look very similar to the general can motors, but the performance is unprecedented. Faulhaber is the most well-known brand

The most wellknown aircoil rotor motor is the Swiss made Faulhaber, available in nearly any size from Märklin Miniclub power alternative to those big G-gauge heavy duty motors. Since it is normally smaller than the standard motor it will replace, there is most likely room for homemade solutions for machining and assembly. Here the motor is simply glued onto the powertruck with a wormgear from the old motor used to distribute power to the wheels.

and they make motors for Märklin Z-gauge engines and up to G-scale motors. My references from now on are though made to the H0-gauge.

PWM OR NOT

The trick to minimise the cons of the iron rotor motors is to feed them with Pulse Width Modulation (PWM), i e short pulses of full operating voltage in order to overcome the friction, but not too much electric energy in order to minimise the "jump forward". By analog operation on a layout this is achieved the easy way via half wave rectifying in the power transformer. That will due to the frequency of

Power transistorer in TO-126 case

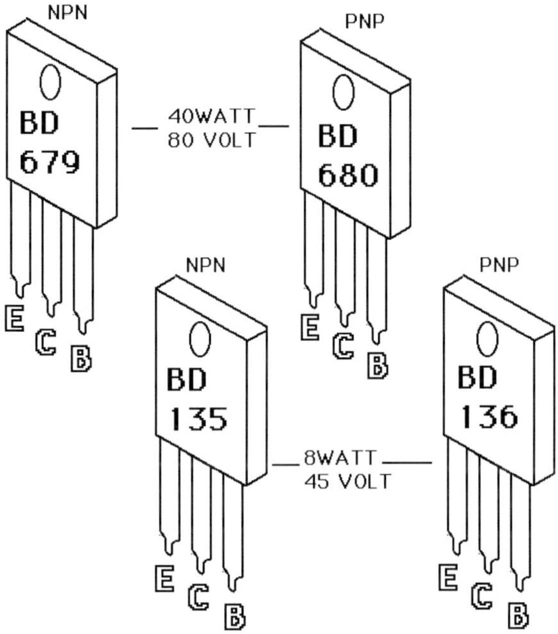

Special-semi-conductors

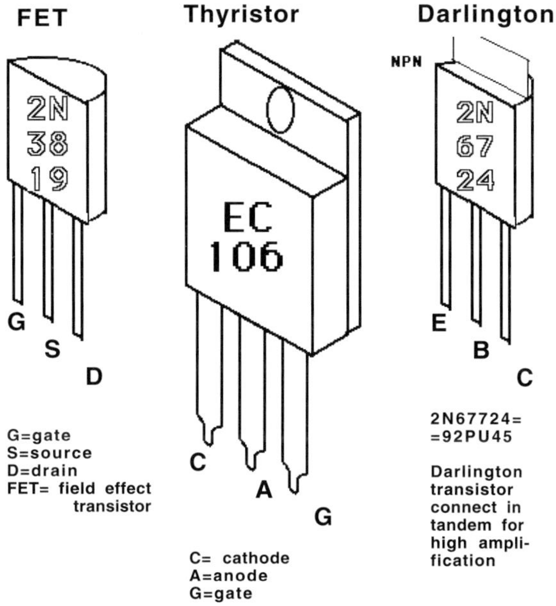

mains AC (in US 60 Hz and in Europe 50 Hz) generate operating voltage with pulses either according to the mains, or the double frequency.

A better way is to generate your own pulses as is done in the power control project in the beginning of this book.

In order to get the most out of the Faulhaber you don't want any sign of PWM at all (there is also one power control project in this book for that purpose). Here we talk of glitchfree and pure DC, which you as an example would find from battery power. The Faulhaber has very high efficiency and power consumption is typically less than 200 mA versus around 1000 mA or even more for the open frame during acceleration.

Further the Faulhaber will accelerate the engine from standstill to low speed in hardly visible steps, and do it absolutely noiseless.

How does the Faulhaber then coop with PWM operation? Well if you can avoid it then do that. Given the above explanation you can't improve the operation further, just wish it is not ruined too much. There is suddenly a clear noise from the motor and if the engine is pulling load (several cars) then heat dissipation from the rotor has no way to leave the interior of the Faulhaber, causing it to get hot and in severe cases overheat. What happens inside is that the glue that holds the air coil windings together melts and the coil gets deformation.

As a summary it is not recommended to run Faulhabers with PWM, especially not when pulling heavy loads. But if you have pure DC it will almost pull anything quietly. And just as an extra advice; most decoders when running DCC generate PWM. There are registers which allow you change the PWM frequency and which will diminish the noise problem, but that is beyond the scope of this book.

The Swiss can motors are getting very popular throughout Europe when powering the high-end locos for peak performance. Two German companies SB-Modellbau and Verbeck have specialized in making repower kits for the most popular engines, also including some US models. This Lima H0 model is normally powered by a very cheap and low performance motor, but since the model itself is unique, it was worth while to invest into the (higher costing) Faulhaber power truck kit which also includes a flywheel for good roll out performance.

Table of Contents